The Thinking Machine

By the same author

Biology and the Social Crisis } *Companion volumes to*
A Natural History of Man } The Thinking Machine

The Thinking Machine

Genes, Brain, Endocrines, and Human Nature

JOHN BRIERLEY

RUTHERFORD • MADISON • TEANECK
FAIRLEIGH DICKINSON UNIVERSITY PRESS

© 1973 by John Brierley. First American edition
published 1973 by Associated University Presses,
Cranbury, New Jersey 08512

Library of Congress Catalogue Card Number: 72-14220

ISBN 0-8386-1364-0

Printed in the United States of America

To the memory of my father
Walter Brierley
and also to his
grandchildren:
Sarah Ann,
Alistair Philip
and
Christopher George

Foreword
by The Rt Revd Hugh Montefiore, Bishop of Kingston

I COULD not read this book unmoved. Dr Brierley explains with ease and simplicity the complex physical mechanisms of genes, brain and glands through which we human beings develop and function. The story is fascinating not only in itself but for its implications. In such a short time modern biology has discovered so much about human beings. How right was the Psalmist to say that we are 'fearfully and wonderfully made', each of us a unique individual but all of us endowed with an amazing complexity of structure and delicacy of balance. How *did* we evolve?

Dr Brierley, as a biologist, writes on the assumption that man is the product of heredity and environment, 'a very complicated determinate mechanism'; in short, a thinking machine. As a scientist he could hardly proceed on any other assumption.

There is of course another side to human nature. Biology by its nature describes the physical basis of personality. But what it means to be a self, my subjective self-awareness as a free, responsible rational cognitive being, capable of making value judgements about truth, beauty and goodness—all this lies outside its terms of reference; and so too does God. An adequate philosophy of life must do equal justice to my subjective experience as a human being and to the objective scientific facts of my physical nature.

Human nature must have this physical basis. I welcome this book warmly because it helps me to understand it.

✠ HUGH MONTEFIORE

Author's Note and Acknowledgements

THIS book is the final one in a trio on the biology of man. All three have attempted to look at man in a scientific way and have been written as a basis for discussion by young students and others interested in the science of mankind. Most of us find man and his ways fascinating, sometimes depressing but always speculation-provoking—'who are we and by what means did we come to be what we are?' are perennial questions posed by writers, poets and scientists and by the young in each new generation. This book, like others before it, by implication poses these questions. Besides the fascination of man as a subject of study, a more pious reason for writing books about man is that they may help people in the long run to care more about individuals, may make them want to improve on the quality of life, and may help them to understand why they sometimes feel and act as they do.

The first book of the trio, *Biology and the Social Crisis*, dealt broadly with genetics and human affairs, and the second, *A Natural History of Man*, with human ecology—population structure, human evolution, the origins of agriculture, the changing patterns of human disease, migration, and war. *The Thinking Machine*, which owes much to V. H. Mottram's classic, *The Physical Basis of Personality*, published many years ago, examines man in the light of our knowledge of genetics, neurobiology, and the endocrine glands. This view of man, with his inherited anatomical, chemical, electrical, and temperamental individuality, plus the fact that, like non-living material, he is made up of the 'dust of the earth' and therefore might be expected to behave with the same regularity and order, requires a counterpoise. Most of us do not order our lives on the basis that each of us is a physico-chemical machine working to a genetic programme. Can it be likely that mental events, as the distinguished biologist Sir Alistair Hardy suggests, while linked to the physical system 'belong to a different order of nature'? The Bishop of Kingston, in the Foreword, adds this other and necessary order and I am most grateful to him for a service quite beyond my range. The poets usually hit the nail on the head, and Emily Dickinson writing in 'A Light Exists in Spring' expresses my opinion exactly:

> A Colour stands abroad
> On Solitary Hills
> That Science cannot overtake,
> But Human Nature feels.

First I should like to thank C. D. Darlington, F.R.S., for his enormous help and encouragement during the writing of this trio of books. I should like to thank too Malcolm Jeeves, Professor of Psychology at St Andrew's University, for reading most of the typescript and leaving it much improved. Dr Michael Casey of the Centre for Human Genetics, Sheffield, gave a similar service and his detailed comments were most valuable. For similar kindnesses and for wise counsel I should like to thank my old friends Mr F. C. Minns and Mr L. C. Comber. Dr Jeffrey Gray of the Institute of Experimental Psychology at Oxford read the sections on brain and endocrines and left them improved. Mrs Janet Carrington of the Department of Biochemistry, Oxford, also saved me from a number of errors and I would like to thank her. I wish I could find a new way of expressing gratitude to Ann who helped most of all.

The book was finished while on a Visiting Research Fellowship at Merton College, Oxford, and I should like to express my gratitude to the Warden and Fellows for their courtesy and help during my term at the College. I am also indebted to the Department of Education and Science for generously granting me sabbatical leave during the period of my Fellowship.

None of the above, however, can be blamed for errors of fact and judgement left in the book.

1973   J.K.B.

Contents

FOREWORD: The Rt Revd Hugh Montefiore, Bishop of
Kingston *page* vii

AUTHOR'S NOTE AND ACKNOWLEDGEMENTS ix

PART 1: The Roots of Personality 1
1. Character and Chromosomes 3
2. More on Heredity and Personality 16
3. Intelligence 24
4. The Parable of the Talents—the Genius and the Gifted 34
5. Men, Women, and IQ 38
6. Class and Brains 46
7. The Intelligence Quotients of Black and White 51
8. Genetics and Education 59
9. Chromosomes, Genes, and Gaols 65
10. Extra Chromosomes and Criminality 71
11. Is Conscience Inherited? 75
12. Angst 89

PART 2: A View of the Brain — 95

13. How Chromosomes Affect the Brain and Intelligence — 97
14. The Brain as a Machine — 108
15. What the Eye Speaks to the Brain — 116
16. Memory and Learning — 127
17. How Emotion is Investigated — 141

PART 3: Chemistry and Temperament — 151

18. The Endocrine Glands: an Introductory Sketch — 153
19. The Chemistry of Character — 172

BIBLIOGRAPHY — 185

INDEX — 189

Part 1: The Roots of Personality

Educability, criminality, and anxiety, are part of the spectrum of human nature and though there are as many human natures as there are men, these three components may help explain the biological roots of the rest.

To understand these roots is to understand a paradox of genetics; that inheritance involves both heredity (like giving rise to like) and uncertainty. This means that the chromosome mechanism which gives the similar stamp between parents and offspring is also part of the same mechanism which shuffles the hereditary units, the genes, to give an unlike offspring: as Plato said, golden parents sometimes have leaden children and a criminal can crop up in a family with no crime record.

The chromosomal foundations to personality are nowhere more clearly displayed than in the brain structure and intelligences of those unfortunate people who inherit chromosomal errors. Even minute errors play havoc with brain and intelligence quotient while errors involving large sections may have extensive effects. An extra Y chromosome, for example, besides predisposing its owner to delinquency under certain conditions, is likely on average to lower his intelligence quotient but no precise figures can be given as yet. Some authorities have established that IQ is pulled down to between 70 and 92 in hospital patients with this abnormality.

Our unique genetic complements, interacting with the unique environment of the particular uterus and with the environment after birth, make each one of us an individual. Brain, nerve, and hormone studies described later in this book confirm that the patterning of these systems is as individual as fingerprints. Such individual characteristics are found in the complex 'wiring' patterns of the brain neurones, in memory stores of the brain, in the 'setting' of the reactivity of the nervous system, in the height of the thresholds over which nerve impulses have to 'climb' at the junction of nerves and in the influence of hormones on the brain long before birth.

Brain studies (in rats) too have disproved the notion of a fixed number of brain neurones. These actually increase in number after birth and are sufficiently plastic to develop unique 'wiring' and perhaps associated chemical patterns in the first two years of life according to the influence of the environment. This confirms again the notion of the tailor-made rather than the mass-produced brain.

If each of us is a machine, which would appear to be a very crude way of putting it, and the evidence points this way, then we are unique machines. These differences between people give the spice to life, but so often individuals are forced into jobs or types of school incompatible with their gifts of personality and intelligence and their talent is wasted.

A famous educational report, the 'Newsom' report, hit the nail on the head when it said: 'There were well over two and three-quarter million boys and girls in maintained secondary schools in 1962, all of them individuals, all different. We must not lose sight of the differences, in trying to discover what they have in common.'

1: Character and Chromosomes

INTRODUCTION

A man's character and temperament provide the set of qualities that give him individuality. The brave, the timid, the zealous, the idle, the stupid, the dull; we all recognize these commonplace characteristics and hosts of others among the people we meet and know and among the larger-than-life characters of literature—the Pickwicks, the Hotspurs, the Falstaffs, the Squeers and Flashmans. We know them too from the past where there have been men whose 'character' has left its mark on history; the blazing personality of Napoleon, who by thirty was master of France, at thirty-six master of Europe, but at fifty-one had died in exile; the outstanding quality of Nelson's leadership, his capacity to win the hearts of plain men; Darwin's passionate curiosity about natural history, which, coupled with his painstaking persistence over detail, led to his hitting on one of the great ideas of science, evolution by natural selection. All these qualities, and many more as we shall see later, are closely linked with the activities of the chromosomes, the microscopic threads of deoxyribonucleic acid (DNA) present in every living cell of the body.

Differences between individuals are readily observed. Also apparent are differences between groups of individuals such as the different races or sexes. As Charles Darwin noted 'the mental characteristics of races are distinct'. 'Everyone', he says, 'who has had the opportunity of comparison, must have been struck with the contrast between the taciturn, and even morose aborigines of South America and the light-hearted, talkative negroes.' Traditionally, men and women are regarded as showing different characteristics of temperament. There is a deep-rooted (but questionable) assumption that man is superior to woman and certainly he is destined to perform a totally different rôle in life. God is regarded as a man; the doctor is usually a man in Britain and America; the vicar is a man; the Prime Minister is usually a man; most geniuses have been men (and most delinquents), and perhaps it is not surprising that authority is naturally invested in the male of the species. Women, in western societies at least, tend to be more passive and responsive, and eager to look after children. They have proportionately more car accidents than men (but fewer serious ones); they occupy a third more beds than men in British mental hospitals and gain more often middle scores in intelligence tests compared with men, who are the highest and lowest scorers. The determination of sex rests on the possession of a portion of a single chromosome.

Enough has been said to emphasize the variety of character, temperament, and intellect that exist among human beings. The question must now be asked: is there a physical basis for such variety? Can we begin to analyse the individuality of men?

First and foremost we must examine whether such characteristics and qualities are an 'award of inheritance', a burning topic indeed, and for this we shall have to know something about the chromosomes, the vehicles of heredity. It is known now that each of the forty-six chromosomes contained in the fertilized egg in the mother's uterus carries a vital code which will help to build a new and unique human being. Not only the colour of his eyes and the diseases he may succumb to, but the quality of his brain tissue, skin, hair, muscle, bone, and glands will have been largely determined at the instant of conception by the coded message contained in each of the forty-six chromosomes. Abnormalities in chromosome numbers lead to abnormalities in the brain. Chromosomes and the genes they carry (our 'genotype') are the very foundations of our life. On these foundations are built unique inherited patterns of biochemical events that are, in some yet obscure way, related to the patterns of an individual's mental activity.

The chromosomes and genes can only operate if the environment supplies the necessary materials and stimuli. The environmental stimuli are important because one of the outstanding features of man is that he depends more on learning than instinct, and owes a great deal of his uniqueness in the animal kingdom to his prolonged childhood and youth during which he is taught and looked after by adults. Yet we must remember that the environment provided by the adults is itself a product of unique genotypes and environments. As we shall see later the quality of the experience may physically influence the brain. We can say that genotype determines our potentialities but environment determines which or how much of these potentialities will be realized during development.

The chromosomes work in ways that are only partly understood, through two agencies which are constantly interlocked: nervous system, especially the brain, and hormones. The brain we might think of as an electrical and chemical system which controls basic movements, vision, emotion, and mood, and the higher functions of reason and creativity. The hormone or chemical system is known to affect our temperament through the effect on the brain of the chemical hormones circulating in the blood. To take a familiar case, about one in five women suffer from moderate, strong, or severe symptoms of irritability, mood swings, depression and/or tension before menstruation. In some cases this leads to crime, poor examination results, lowered school or work performance, and decreased efficiency generally.

We shall need to devote a few chapters to all these ideas but we may sum up the basic message of this book: that a good deal of what we put down as 'character' and 'personality' is the interaction between unique genotype and unique environment, not only the environment of life but the pre-life of the uterus, itself unique. It is this interlocking of genetic uniqueness with unique environment which gives rise to the vast range of human nature.

What is important to the scientist investigating human nature is the hypothesis that observable events have material causes, and it is his job to discover these causes. Indeed this is the only working hypothesis he as a scientist can use. The facts set out in this book do show how surprisingly far this hypothesis has already carried the investigators, especially those engaged in neurobiology.

Some of the improved knowledge of the controlling systems of behaviour can be used deliberately by each one of us by the application of will and conation to improve human happiness, reduce misery, and enlarge experience. Other pieces of knowledge can be used by doctors to control or cure mental illness. The more we know about ourselves the more we can do to help ourselves and others. But to govern our lives and our striving as ordinary people facing the day to day problems of life, we are compelled to adopt another working hypothesis: we need to believe in the presence of a 'joker' in the pack in the form of free-will. Indeed a large section of mankind believes as well in a 'master-card' in the form of a deity. One or both hypotheses are imperative for healthy functioning of mind. If we believed that we were simply propelled by machine-like processes to a programme laid down by the genes, much of meaning would drop out of life. None of us, of course, thinks like this, but what this book attempts to do is to describe some of the physical machinery that might lie behind human nature.

What emerges, as expected, is no comprehensive theory of human nature. What we have is a great mass of often uncoordinated facts from genetics, neurobiology, endocrinology, and psychology about mental events and how they may affect the body and the body the mind, but facts that are intrinsically interesting, suggestive, and speculation-provoking. They prompt the questions; 'who are we, and how did we come to be what we are?'

This book begins its description of the basis of human nature with evidence from the chromosomes, moves on to the brain and nervous system, and finishes with endocrines.

CHROMOSOMES AND GENES

The fertilized egg, a single cell, from which each one of us developed, can just be seen by the naked eye and is smaller than the stop at the end of this sentence. Indeed all the eggs from which the population of England were derived could be packed inside a hen's egg. Half the population of England could be accounted for by the amount of sperm in a globule the size of a pin-head. Only one of the millions of sperms released in one ejaculation is necessary to fertilize the egg, and once it is fertilized other sperms are blocked from entry. The human sperm and egg both normally carry a package of twenty-three chromosomes which form the genetic dowry of the child from each parent. The chromosomes are microscopic, thread-like bodies, each of which carries a code for the future development of the individual. (See Figure 1.)

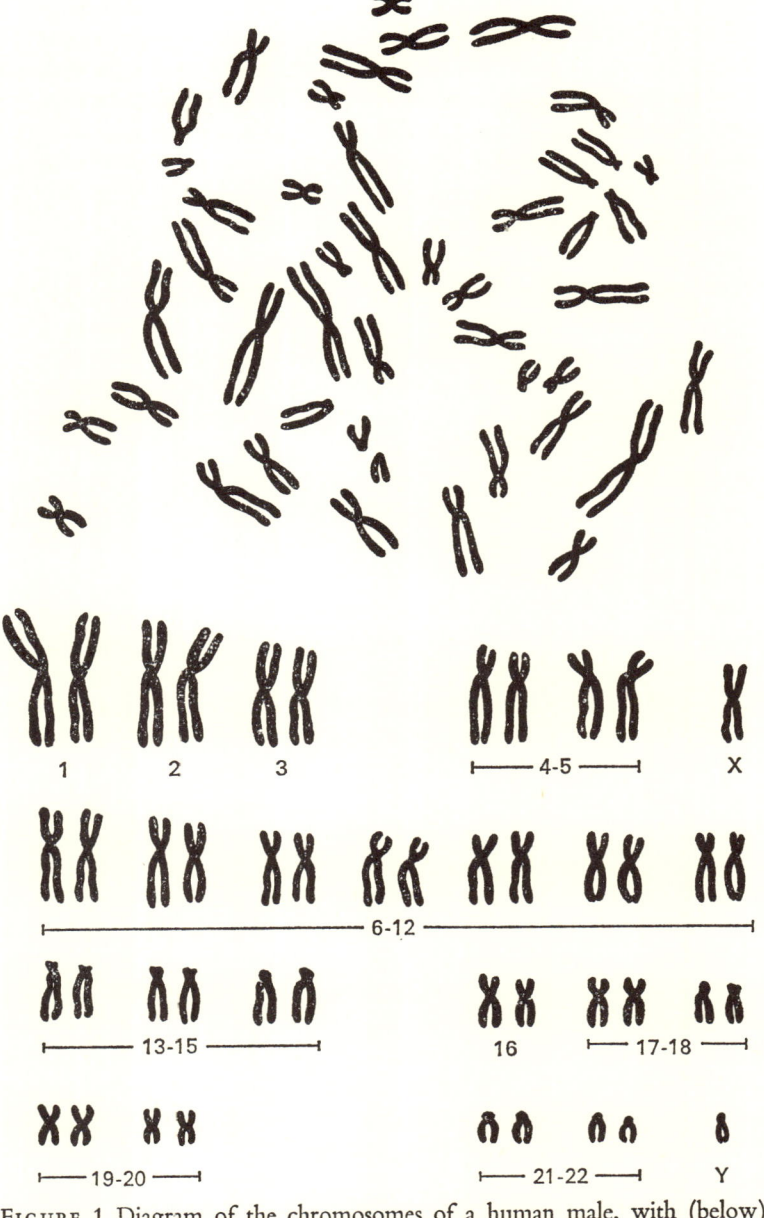

FIGURE 1 Diagram of the chromosomes of a human male, with (below) identifications. Note that all except the Y chromosome are in matching pairs.

The code is contained in units called *genes* which are so small they cannot be detected by the microscope. These remarkable gene-units (and the number of major genes in man is probably 10^4 or 10^5 on the twenty-three pairs of chromosomes) contain the chemical instructions for the design and development of nose, eyes, mouth, brain, and so on. The primary functional unit of heredity is the gene. Each gene occupies a fixed place on a chromosome and is passed on to the next generation through the sex cells.

The fertilized egg, then, has forty-six chromosomes in twenty-three pairs, and one of each pair has been obtained from the father and the other from the mother. Forty-four of these chromosomes can be seen under the microscope to be in matching pairs. But in a male there is a discrepancy in the twenty-third pair. One chromosome of this pair is much smaller than the other and is known as the Y or male sex chromosome. Its partner is called the X chromosome. A male, then, has twenty-two matching pairs of chromosomes, the autosomes or non-sex chromosomes, and an odd pair XY. (See Figure 1.) The female has no such odd pair and has twenty-three pairs of matched chromosomes, of which one pair is the XX pair of chromosomes, the rest are the autosomes. We shall return to the determination of sex shortly. Like the chromosomes which bear them, the genes are always in pairs, one of the pair having been obtained from the mother, the other from the father. If the person has an identical pair of genes which each produce the same chemical instructions he is said to be *homozygous* for that gene character; one who carries an unlike pair is said to be *heterozygous*. When the gene pair is the same no conflict arises, as the geneticist L. S. Penrose puts it. But if they differ some decision must be taken as to which instruction is to be followed and which ignored. Generally speaking, one instruction is suppressed by the assertive or *dominant* gene. The suppressed or less forceful gene is called the *recessive* gene. Sometimes, and possibly very often, both instructions are obeyed and the results of both genes show in the character which develops. It is, of course, rather loose to speak of a gene being 'dominant' or 'recessive', for dominance and recessiveness refer to the *character* or end result rather than to the gene.

Very roughly, inherited defects in people can arise through three agencies:

(i) Possession of one 'bad' dominant *gene* or two 'bad' recessive genes. Abnormalities arising from these agencies are rare; of the order of one in 10 000 of the population, or less.

(ii) Abnormality in the numbers or types of the *chromosomes* so that the harmonious balance of the gene outfit on these chromosomes is disturbed. The frequency of these chromosome abnormalities is of the order of one in a 1000 of the population.

(iii) Possession of a combination of genes which happen not to work well together and incompatibility of genes and environment in some people. This third category of abnormality due to the possession of many genes of small effect (multifactorial inheritance) is much more common than (i) and (ii). Indeed many disease-states with this basis have a frequency of one in 100 in a population.

We shall return to these categories later.

When the fertilized egg in the womb begins to divide into two, four, eight, sixteen cells, each cell normally contains the set of forty-six chromosomes. The constancy of the chromosome number is due to the fact that each of the chromosomes, which at certain stages is a double thread joined at one point, is divided lengthwise into halves which are exactly equal. The two single threads of each chromosome are attracted by some force to opposite ends of the cell to give two new groups of chromosomes each containing the original number of forty-six. Shortly after this 'polarization' of the chromosomes, the cell divides into two. (See Figure 2.) Before the process of cell division can be repeated, each single chromosome thread exactly reproduces itself and becomes double. This process of cell division is called mitosis. When a baby is born it has about twenty-five million million cells all identical in chromosome number and exactly duplicating the genetic information in the original fertilized egg.

THE UNIQUENESS OF THE INDIVIDUAL

We now turn to a fundamental question related to the chromosomes. Why is it that a man (or woman) is born *unique* with no one, quite like him in body or mind? Briefly, it is because the sex cells, the egg and sperm, contain unique combinations of genetic information on the chromosomes. In fact the probability of one sperm or egg having the same assortment of parental chromosomes as another is one in 8 388 608. To put it another way, each human parent has the potentiality of

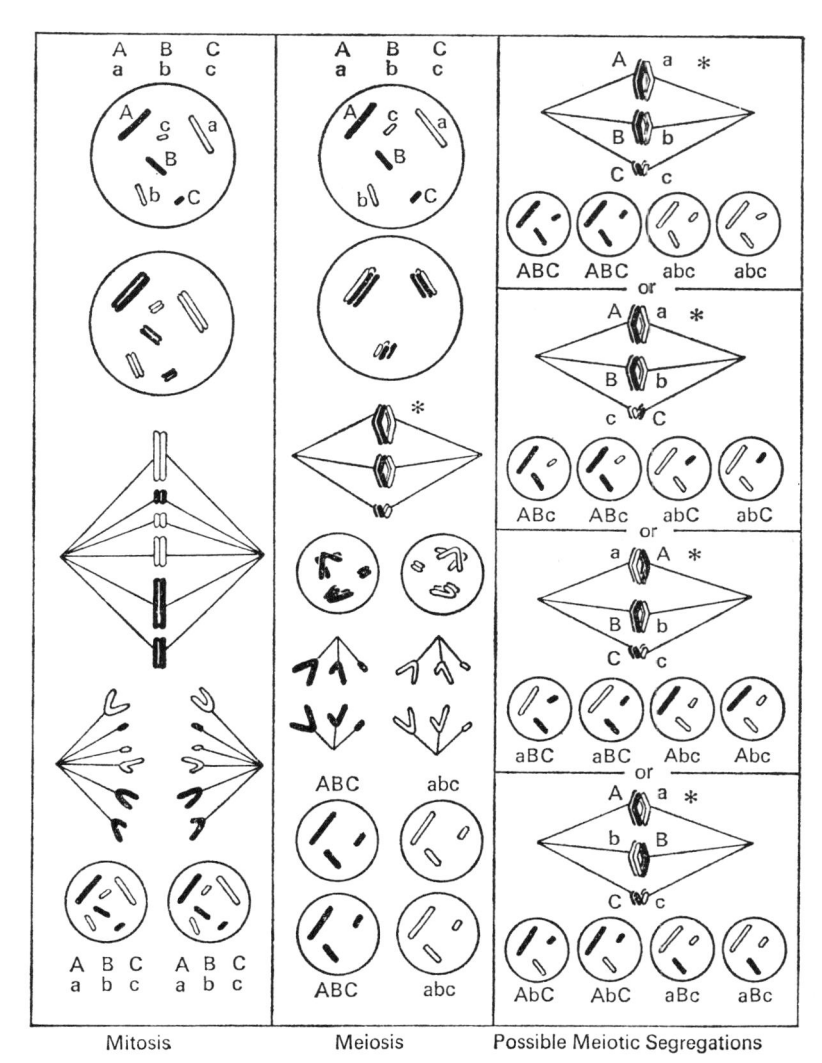

FIGURE 2 The mechanism of mitosis and meiosis. Three pairs of chromosomes only are shown each pair carrying a pair of genes: Aa, Bb, Cc. In mitosis the products of the division are identical with each other and with the parent cell. In meiosis the gametes (eggs or sperms) end up with half the number of chromosomes of the parent cell and also, as a result of: (a) crossing over where the genes are swapped between partner chromosomes as shown in Figure 3 and (b) the way the chromosome pairs align before they separate in the first meoitic division, each gamete is very different. The first meiotic division is marked by an asterisk. Eight kinds of gamete in equal proportions are shown in the figure due to the chance alignment of chromosomes. Thus A, B, and C and a, b, and c might end up together as in the top right-hand diagram. But A, b, C, and a, B, c might also be formed as in the bottom right-hand diagram. (From Srb, Owen and Edgar, *General Genetics,* Freeman 1965.)

producing over eight million unique types of sperms if a man, or eggs if a woman. How does this come about?

If the sex cells were formed by mitosis as described above, a damaging situation would arise at fertilization because a sperm and egg fusing would together contain ninety-two chromosomes instead of forty-six. To stop this multiplication process and also to produce individuals and not standardized human beings, certain cells in the ovary and testes are 'programmed' to produce eggs and sperms containing half the normal number of chromosomes: twenty-three instead of forty-six. This process is called meiosis (see Figure 2) and is basically similar to mitosis, although it differs in several significant ways. At first the chromosomes pair up but do not divide. Each chromosome, though connected at one point, is double or in the process of doubling so that four strands are associated in a tetrad. One of each pair is then pulled to opposite poles of the cell so the number of chromosomes is reduced to twenty-three in each cell, for the original cell has now divided into two. This is the first stage of meiosis. These two cells now divide by ordinary mitosis and so the original cell gives rise to four reproductive cells with twenty-three chromosomes in each. Each chromosome doubles itself so that at the end of the second stage of meiosis there are four cells each containing twenty-three double-threaded chromosomes. (See Figure 2 where only three pairs of chromosomes are shown for simplicity.) This 'reduction' division thus stops the doubling of chromosome numbers in the fertilized egg. But this is only half the problem.

How are duplicate human beings prevented from arising? Basically this is due to a considerable reshuffling of the genes during the first stage of meiosis. When pairs of chromosomes come together at the tetrad stage, each chromosome is a double thread held together at one special point, the *centromere*. Whole sequences of genes along the chromosome threads change their allegiance and swap places with those on the threads of the partner chromosome. Physically there must be a break at identical points in the two threads so that an exchange is effected and whole blocks of genes are switched (crossing-over). The threads then connect up again though now with different gene arrangements. So when each pair of chromosomes is pulled to opposite sides of the cell as previously described, the arrangements of genes on the chromosomes will be quite different from the arrangements the chromosomes started with. And in the second division of meiosis, which ends up with four cells each containing twenty-three chromosomes, each cell is bound to contain quite different combinations of genes along the chromosomes. (See Figures 2 and 3.)

This extraordinary genetic mill thus grinds out a constant stream of novel combinations of chromosomes and rearrangements of genes along them. Each sperm or egg carries a unique combination and this, interacting with unique environment, leads to unique individuals. A single human couple therefore has the potentiality of producing a vast number of differing types of children.

As soon as sperm and egg combine their chromosome packages, that is within half an hour of the sperm reaching the egg, all the myriad of traits and potentialities in the body are decided: the broad stamp of the human species, the racial features, the family face or not, blood groups, the tendency to be tall or short, fat or thin, brainy or stupid, healthy or prone to disease. With the thousands of genes involved, not

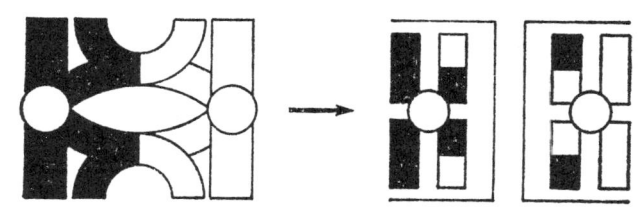

FIGURE 3 The sequence in which a white chromosome becomes entwined with a black one, creating a pair in which the genes are mixed at random points.

only does like give rise to like but, because of the reshuffling of chromosomes and genes in meiosis and crossing-over, there is the vast possibility of novelty in existing family patterns.

One matter amongst many others, of a basic kind (that is, those relatively independent of the environment) is settled in the first half-hour of life; the sex of the child. There is nothing mysterious about this, for male or female child is determined by the father's sperm. Half a man's sperm cells carry a Y chromosome and half an X. The Y chromosome is the male determiner so if a Y carrying sperm fuses with an egg, an XY egg is formed since eggs each carry one X chromosome. The XY egg eventually makes a male child. If, on the other hand, an X carrying sperm fuses with an egg, the egg will be XX and a female child will be born. In theory males and females have an equal chance of birth but many more male than female embryos start development, though many of the former are discarded by the womb.

In the end, in the United States for example, 106 boys are born for 100 girls, in Cuba 101 males for 100 females, and in Greece 113 males for 100 females.

Why more boy than girl babies are conceived is unknown. Perhaps the Y-carrying sperm is lighter than the X-carrying one because of the smaller size of the Y chromosome as compared with the X, and thus gets to the egg more quickly, but no firm evidence supports this notion. While the X or Y chromosome of the sperm acts as a switch leading to formation of ovaries or of testes, the division between man and woman is not a sharp one. In the secondary sexual characters of man and woman—depth of voice, distribution of hair and fat, aggressiveness or timidity—there is a continuum between man and woman which is probably due to variations in the genes, not just on the XY chromosomes, but on all forty-six. We must not forget that besides the unique qualities and potentialities determined by the genes, a child at birth has already at this stage been in the unique environment of its mother's uterus. As Thoday points out, the differences at birth are not only a consequence of the genetic variation of individuals but may have been contributed to by environmental factors, some of them determined by the genetic make-up of the mother as, for example, the environment of the uterus.

HEREDITY AND UNCERTAINTY

The raw material of our unique make-up of body and mind is then, based on the genetic information on the chromosomes derived from the two parents. This mechanism ensures heredity: the similar stamp in the properties of parent and offspring passed down the generations to ensure, as Thomas Hardy wrote, 'the family face ... the years-heired feature that can in curve and voice and eye despise the human span of durance'. But it also ensures 'uncertainty' because of the continual taking-apart and rearrangement of the gene outfits in meiosis, as we have seen. We are all familiar with the very bright individual in families where both the economic and intellectual status of the parents would, one might think, damn every child they had to hopeless failure. Bunyan the tinker, Kepler the drunken innkeeper's son, Faraday the blacksmith's son, and Dickens the boy working in a blacking factory, confound the simple notion of heredity. Heredity involves both 'heredity' and 'uncertainty'. In other words heredity involves both the handing on of similarities and the introduction of new characteristics.

Another source of uncertainty is when a gene suddenly changes in the sperm or egg cell. Such a change or *mutation* is a rare event and almost always it is a change for the worse. It might affect things like eye colour or blood group in the offspring or give rise to a disease like fibrocystic disease of the pancreas.

Fortunately, because genes are always inherited in pairs (one from each parent on the partner chromosomes), most but not all mutations are hidden or recessive in the offspring. That is, the offspring is said to be a 'carrier', technically a heterozygote. In order to express themselves, these recessive 'bad' genes, to put it crudely, must be present in double dose, one mutation passed on in the sperm, the other in the egg, not just in one or the other. Why do we call them 'bad' genes? For three thousand million years or so selection has been favouring those genes which best confer fitness (ability to leave offspring); mutation will generally produce genes which give a lowered fitness, a change for the worse, as indicated above.

Mutations seem to happen spontaneously or be caused, for example, by serious damage from radioactivity. Most of us carry at least two or three quite serious recessive genes but fortunately most of us are lucky enough to marry someone who carries two or three quite normal (that is dominant) ones affecting the same character. More than 2000 diseases are known to be triggered by 'bad' genes, but the number of diseases to which some people may be genetically disposed may be many more.

HEREDITY AND BACKGROUND

We must now ask the question, how is heredity—'all that a man brings with himself into the world', as F. Galton put it—affected by the influences of the world? This is a matter which we shall have to explore further, but it is not a simple one. We may say, however, that the upper and lower boundaries of an individual's capabilities and performance in a particular quality like intelligence are set by his unique combination of genes. And indeed the unique environment provided by the mother's uterus might also influence intelligence quotient (see Chapter 3) and other qualities. Common sense and experience suggests too that books, conversation, a good or bad home, school or health or nutrition will influence performance and behaviour. Many a sharp schoolboy, as Sir Cyril Burt the psychologist tells us, fails at school because he has come to look on the world as a kind of fun fair. So often lack of stamina and grip turns the quick mettle of the schoolboy into the 'blunt fellow' of middle age. But it is quite likely that heredity has had something to do with the lack of grip or failure in health.

Where a very bright boy is discovered in a family of low economic status it often turns out that the family has unusual characterisics which gave the youth an advantage—thrift and ambition, or an interest on the part of the father or mother in literature, art, or science. D. H. Lawrence and his mother's influence come to mind, and it is quite likely that her ambition and artistic gifts were inherited. Nature and nurture co-operate and we cannot say exactly how much has been contributed by either.

We should be very wary indeed, however, of the simple notion that environment 'shapes and moulds' us like plasticine and that our heredity allows it. Heredity is grittier than this. Witness the conflict between parent and child. Even though the child has a better chance of being like his parent than anyone else in the world, the genetic individuality of the child overrides the parental environment. Heredity here, as in many cases, seems to 'overcome' environment.

By this is meant that each one of us does not 'respond' to the environment like a plant thrusting to the light. Rather he may choose instinctively the environments that fit him in body and mind; that is, we choose particular environments that fit our diverse and unique heredities. Consider an ordinary family: one will go for a walk for exercise and see nothing; another will walk with another and discuss economics or dog racing; another will root in the hedgerows for plants; and yet another will choose not to go for a walk because he prefers to sit at home and read a book.

Investigations have shown too that the muscular, outgoing man is more often found in jobs that involve some rough and tumble—the army or physical education or crime! The bean-pole shaped introvert gets himself into quieter jobs such as on the research side of a factory. We shall return to the environment-heredity story later but we may say that heredity provides the basis of variety. Different and powerful heredities, such as those of the genius or criminal, will often create environments to suit themselves, but most of us seem to slot into the niche that suits us best.

2: More on Heredity and Personality

TWINS

MOST mothers of twins do not know (or care) that their offspring provide one of the keys to the understanding of the inheritance of character, temperament, intelligence, and health. Twins of both types can help us to know more about ourselves, and indeed help us to estimate how far a criminal, homosexual, alcoholic, or genius is 'born' or 'made'.

Two-egg or non-identical twins are scarcely more alike than ordinary brothers and sisters; one-egg or identical twins are very often closely alike in looks, blood groups, and even finger-prints and are always alike in sex. Two-egg twins are derived from two separate eggs, each of which has been fertilized by a different sperm, and thus lead to quite different people since they have quite different combinations of chromosomes. One-egg twins are derived from the splitting of a *single* fertilized egg and thus they have facsimile packages of chromosomes and are likely to be similar.

How can twins help us to measure the relative effect of heredity and environment? Since one-egg twins have identical heredities they can be used to show the kinds and degrees of difference in health, educability, and personality which arise mainly from the environment. And when one-egg twins are separated at birth, as sometimes happens when they are orphaned, the effects of the environment can be estimated even more sharply, or to put it another way, the force of genetic determination can be assessed more clearly. Two-egg twins can be used to show the shades of difference that arise within a family by heredity. Since the family environment will be reasonably similar for all, the play of heredity will be exposed more sharply.

The picture of the interplay of heredity and environment given by twins is not as clear-cut as this however. Perhaps 10 per cent of one-egg twins differ considerably in such properties as birth weight, temperament, and general intelligence. This is possibly because of gene mutation in one egg only, after the splitting of the fertilized egg into two. But more important, unequal division of the cytoplasm (which surrounds the nucleus containing the chromosomes) when the fertilized egg splits, can lead to such differences, and differences can arise from the different positions in the uterus and unequal blood supplies.

Cytoplasmic differences do not occur in two-egg twins and it is for this reason that two-egg twins are more alike in birth weight than identical twins. Despite these anomalies, however, twins together with other hereditary studies have provided us with a powerful instrument by which to examine the basis of individuality.

What other observations from heredity can help us?

FAMILY TREES

First, there is the study of family trees which show that individuals and families differ in their character, health, and abilities and that the way they differ can be explained by the fact that relatives have in common something determining these qualities—chromosomes and genes. We shall be referring to family trees in later chapters, but the work of Galton on *Hereditary Genius*, published in 1896, illustrates their value. Galton tried to understand man in a new way. He argued that if men are observed as dispassionately as animals and if their varied talents and abilities of mind and body are measured, and if they can be followed through life as individuals and in families from generation to generation, then some idea of the transmission of ability can be gained.

Contrary to the popular idea that genius often appears out of the blue (it can do, of course, as we have said earlier, by an advantageous recombination of parental genes in the offspring), outstanding talent is generally built up step by step, generation after generation (see Chapter 4). In time a new outstanding gene combination is put together and a Bach, a Darwin, or a Winston Churchill appears. According to the type of marriage, Galton argued, 'there is a regular increase of ability in the generations that precede its culmination and then a regular decrease'. Instead of clogs to clogs in three generations it is average to average in seven or eight. Thus when Galton examines his outstanding men from all walks of life, the chance of his finding other outstanding men decreases as he moves away, in terms of relationship, from the one he begins with.

Take the case of the outstanding judges of England between 1660 and 1865, studied by Galton. Only 26 per cent of the fathers of his distinguished judges were distinguished, 36 per cent of their sons, 7.5 per cent of their grandfathers, 0.5 per cent of their great grandfathers; and similarly for their descendants. The effects of a particular heredity are dispersed as the outstanding combination of genes in our eminent man are scattered in new arrangements.

There was one important point Galton missed, however, yet it was clearly demonstrated in his own work. He did not realize that *assortative mating* or like marrying like is to a large extent important in maintaining high or low intelligence or this or that ability, and this matters in the step-by-step building up of genius. Assortative marriage is one of the foundations of society because it fosters marriages between people with similar interests and abilities. And as far as temperament is concerned there is some evidence that there is an attraction between people of similar psychological character.

INTELLIGENCE TESTS

Second, besides family trees and type of marriage, educational experience through intelligence testing can help to fill in the hereditary story. Burt tells us that gifted children from three widely different parts of the world differ from the average child to the greatest degree in the intellectual and volitional, and least in the moral and social characteristics. They resemble each other, Burt suggests, in those characteristics which may be assumed to have a hereditary basis and differ from each other chiefly in those qualities most strongly affected by environmental and social conditions. We shall be referring to the measurement of intelligence in Chapter 3, but basically it was Galton's work—the idea of a *scale* of merit, a one-dimensional standard of variation—that gave rise to the intelligence quotient or IQ test.

IQ tests are useful in the grading of unclassified individuals in the middle ranges of intelligence. A usual way of scoring IQ is to assign to the child a mental age, which is the age at which the average child makes the same score. IQ is then obtained by dividing the child's mental age by his real age and multiplying by 100. Thus a child with real age six years who scores as high as the average child of nine would have an IQ of 150. One who scored as low as the average child of three would have an IQ of 50. IQ study is a vital step in the study of the inheritance of 'intelligence'. But it is a first step. Such tests should lead to the study of temperamental qualities, important too for success in school achievement, but in practice they usually exclude it. Indeed, often the tests perpetuate the myth that people are composed of discrete, separable, measurable factors like 'intelligence' when we all know that the way we feel affects our thinking.

ORPHANS

The study of foster, orphanage, and illegitimate children is a third attack on the heredity-environment problem and one we shall need to examine later in Chapter 3. Will the children of the natural parents who become mentally ill, or take to crime, be inclined to mental illness or criminality even in the homes of healthy, normal, foster parents? If so, heredity provides a powerful imprint. We know too that some illegitimate children of high ability have a father well above the mother's own social and intellectual status. The father, just a casual acquaintance of the mother, has taken no further interest in the child and the child has never seen him. In such a case it is quite out of the question to attribute the child's high intelligence to the cultural opportunities of the home.

Then, will adopted children do better or worse in IQ tests than the natural-born children of foster parents? If markedly different then heredity is overwhelming environment. And will the IQ of orphanage children, despite the uniform environment of an orphanage, reflect the social class from which they came? This question implies that the genetic composition of the various social groups (see Chapter 6) is different, and IQ studies reveal just this. If our orphans reflect their different hereditary backgrounds, and they appear to do so, then a social class is indelibly imprinted on them from the moment of conception.

BAD GENES

Fourth, is the fact that mental deficiency, IQ of 70 and below (normal IQ is 100), is of two kinds; one sharp and clear-cut to produce idiots (IQ of 20 and below) and imbeciles (IQ of 50 to 20); the other, which seems to be the shading off of the normal spectrum of intelligence, from IQ 70 to IQ of about 45.

The first type, idiots and imbeciles, are in some cases the result of single recessive genes inherited in double dose, as in phenylketonuria referred to in the next paragraph, and to chromosome errors described in Chapter 13. (Other causes of idiocy and feeble-mindedness are the results of injuries to the brain at birth.) The second type of mental defect, feeble-mindedness, appears to depend on many genes of additive effect. Such multifactorial inheritance, to which we shall refer later, make large or small contributions to such important human characters as intelligence, height, and fertility. (See Chapter 3.) For 'intelligence' (and the other qualities) they act together independently and cumulatively to make the total score of IQ. Fraser Roberts has shown that many cases of serious feeble-mindedness defined by IQ scores in the range from 45–70, are probably due to a high concentration of such genes for low intelligence being inherited from parents of only slightly lower than average intelligence themselves. Assortative marriage coupled with the chance recombination of genes at meiosis sometimes deals out unfortunate heredities.

There is the fact, too, that a single defective gene, inherited in double dose, one from each healthy, normal parent, can produce deep-seated, manifold mental consequences as in the disease phenylketonuria where the diseased baby is rapidly on the way to becoming an imbecile. Indeed early studies (G. A. Jervis) showed that the great majority of patients had IQs of less than 20 and nearly all the remainder between 20 and 50. The defective gene, like many others now known, causes errors in normal body chemistry, and a striking feature of the disease is the presence of large quantities of phenylpyruvic acid and other unusual substances in the urine. This is because a normal constituent of some foods, phenylalanine, cannot be properly metabolized by the affected babies and the by-products of this spill over into the urine. The failure to cope with phenylalanine is due to the baby's inability to produce a critical chemical compound, an enzyme which is necessary to digest and utilize phenylalanine. In phenylketonurics both copies of the (normal) dominant gene which 'knows how' to make this enzyme are lacking and replaced by a faulty (recessive) pair, which are 'ignorant'. Something like 30 milligrammes of phenylalanine per 100 cubic centimetres of blood plasma pulls down the IQ score to less than 20 in these unfortunate people. But the IQs of phenylketonurics vary; a few have been described by S. Coates with normal IQ (95–107). The condition can now be recognized at birth and can be controlled by special diets, almost free of phenylalanine.

Discoveries like this have revolutionized ideas about the causes of mental deficiency and also the ways in which genes work. Phenylketonuria is probably responsible for about 0.6 per cent of the mentally defective people of the world who are in mental institutions. Previous to this doctors had thought that all backwardness was due simply to faulty development of the brain and not, as is really the case, sometimes due to the body's failure to make an enzyme necessary for a particular chemical job. The mental backwardness is thought to be due to poisoning of the nervous system by abnormal chemical substances (phenylalanine and its by-products) accumulating in the tissues. It is salutary to remind ourselves that much of what passes for normal behaviour and conduct depends on having the right genes and the right enzymes to do the necessary chemical jobs of the body and brain. This is a subject to which we shall return in later chapters.

MENTAL ILLNESS

Fifth, hereditary studies of mental illness with one-egg and two-egg twins, and studies on the brothers and sisters of mentally afflicted individuals, show that there is a genetic component to at least some of these conditions. Eliot Slater showed, for instance, that chronic depressive mania is twenty-five times commoner among the brothers and sisters of afflicted individuals than in the general population, and there is evidence that some mental illness is caused by faulty chemical balance in the brain. This is why some mental illness can be controlled by drugs.

CHROMOSOME MISTAKES

Sixth, chromosome mistakes at fertilization can give rise to a number of abnormalities in looks, behaviour, and brain structure (see Chapter 13). Indeed defects in every system of the body can usually be found if looked for in sufficient detail in patients with chromosome abnormality.

An extra, non-sex, chromosome is responsible for 'mongol' children in whom each cell, instead of the normal 46 chromosomes, contains 47; the formula for a mongol male would be $45 + XY = 47$, $45 + XX = 47$ in a female. This one extra chromosome, making a trio of similar chromosomes instead of a normal pair, leads to slight anomalies in nose, eyes, mouth, ears, and shape of head. A mongol baby is born to about one in every 666 mothers. Besides being physically dwarfed, mongols are seriously mentally retarded, their brains are abnormal in structure and they are exceptionally susceptible to infectious diseases. Yet their disposition is usually most happy.

An extra Y chromosome, giving men with $44 + XYY = 47$ chromosomes, is associated with very tall men, sometimes with violent, aggressive behaviour. This is one of the most dramatic links between chromosomes and personality. Incorrect chromosome balance in both sex and non-sex chromosomes can therefore lead to serious disturbances of growth, looks, intelligence, temperament, and brain structure.

HEALTH AND LOOKS

Seventh, although looks and health are mainly outside the scope of this book, it is commonplace that both will affect behaviour and personality. The pretty blonde with the good legs has no shortage of admirers and this is bound to affect personality. So will short sight or haemophilia or duodenal ulcer. All of these conditions affect personality, yet genes affecting length of eye-ball, blood clotting, or the tendency to develop ulcers are not usually thought of as 'behaviour' genes. None the less, health and looks are intimately bound up with behaviour and personality and come within the scope of our argument.

ANIMAL EXPERIMENTS

Finally, to add to our knowledge of the genetic origins of human individuality we must include indirect evidence from experiments on the behaviour and temperament of animals. The proper study of mankind is not, indeed, rats or dogs. However, dogs in particular have provided a useful instrument through which we might understand more about ourselves, particularly our temperament. Every breed of dog has its own instinctive character—the kind, gentle labrador, the aggressive wire-haired terrier, the enthusiastic, gay-spirited Irish setter, the submissive cocker spaniel. Of course it is not the dog's mind that is much like our own; it is the variety of his mind that is the useful tool to the studies of man. In behaviour, instinct, temperament, the dog varies about as much as the races of man; in intelligence or educability he varies much less. Nevertheless, he is more suitable for comparison with the races of man than the races of any other animal. Dog-breeding then, with carefully designed experiments, can throw light on the importance of the mechanism of heredity on variations in human temperament, character, and educability. Collies, Alsatians, and poodles are more fearful than boxers and Scotch terriers of strange but harmless objects. Aggressive biters of people are guard and sheepdog types, though bulldogs and mastiffs are gentler despite their looks. Shyness runs in families and breeds of dog. Wildness, tameness, cowardice, shyness, learning ability, lethargy, and activity, are all qualities that can be crossed and are thus under genetic control.

All these various strands of information can be woven into a convincing total picture which shows that heredity is a powerful element of human character, temperament, personality, call it what you will. But as we have stressed earlier, the genotype determines only the potentialities of an individual and the environment determines which or how much of these potentialities shall be realized during development. As we have hinted earlier, however, on page 15, the interaction of heredity with environment is far from a simple response. One ingredient in a very complex situation is the instinctive choice by an individual of the environmental niche that suits his unique heredity.

In the next chapter we shall consider one important dimension of personality—intelligence.

3: Intelligence

WHEN Gulliver found himself in Lilliput he was a giant, but in Brobdingnag he was dwarf; in both countries he was quite out of context and could not be thought of on the same scale as the inhabitants of either society. Although this observation may seem irrelevant to a discussion on heredity and intelligence, it has point because 'intelligence', like Gulliver's height, is an abstraction, and if we give it meaning more than as a mere relationship we do so at our peril. Moreover we simply cannot measure the heritability of intelligence once and for all. All that can be made is an estimate of the inborn nature of intelligence as measured from scores in a particular test by a particular group of people living at a particular time in a particular area. If any one of these were different, the hereditability might be different too.

Intelligence then, like height or blood pressure or fertility, is only a relative quality dependent very much on the place, time, and culture. We can say nothing about an individual's measured IQ as such, except in relation to other individuals in a population. To show how 'relative' a quality intelligence is, if a society decided to bring up its children under absolutely identical conditions, the environmental factor would decline and the dependence of intelligence on heredity in a century might approach 100 per cent. There are four other general points that perhaps need to be made about intelligence before we analyse its link with heredity.

First, 'intelligence' cannot be isolated from other facets of human personality; in any case, of course, it is not a definite unitary entity depending on particular genes, as we shall see. In fact it is pertinent to the theme of this book to demolish the notion that people are composed of bits that can be separated out and analysed independently.

Most of us would agree that intelligent activity is the ability to grasp the essentials of a situation and respond appropriately to them: and it is commonplace to say that the 'activity' can be affected by mood, attitude, temperament, character, or aspiration. Put simply, thinking and feeling cannot be separated. Miss X says of her boss, 'he gets so worked up that he can't think straight' (strong emotion clouds clear thinking), or Mrs Y to a neighbour about her son's eleven-plus failure, 'the teachers say he's clever but he's not interested in school work' (insufficient motivation lowering functional intelligence), or Mr Z, 'I prepare my talk in minute detail but when I get there I seem to go to pieces and it falls flat' (anxiety reducing performance). As Galton recognized and described so clearly over ninety years ago, a person's success in all walks of life depends on at least three types of mental quality: 'he must possess the requisite abilities; he must respond with eagerness and zeal; and he must sustain the necessary effort'. In the jargon, success depends on three types of mental quality—cognitive, affective, and conative.

Second, we have said earlier that IQ is defined rather narrowly, as the quality measured by the highly fallible instrument of intelligence tests (the innate, general, cognitive ability of Burt); orthodox tests, not of the type which tests 'creativity'. We now know, of course, that intelligence is of many kinds, a name for a group of overlapping skills, and that it has to be measured not on one scale but many. It is probably fair to say, though, that the administration of a well-established intelligence test is still the most reliable way of assessing an individual's capacity for intelligent behaviour. The tests seem to assess the ability to spot relationships and apply them in new situations, and this ability underlies the mental effort necessary for all sorts of processes. The data referred to later in this chapter are based on such IQ tests. This is not the place to go into the different kinds of IQ tests, but estimates of the heritability of intelligence by tests vary according to whether the test is dependent on an acquired knowledge and facility in the handling of words. Clearly a test stripped of such requirements is a better test of heritability.

Third, there is the question of racial intelligence, a very hot topic, on which we shall comment in Chapter 7. The tests of IQ mentioned above would be of little use with Arabs or Red Indians, for example, because they involve knowledge and concepts that belong to our own type of society, besides of course requiring knowledge of a particular language. In fact, the white Westerner in doing such tests is playing on home ground, the Arab, Red Indian, Negro, or Japanese is playing away. Even non-verbal tests which ask the person to arrange coloured blocks, to draw a man, to trace a path through a maze, and are devised to overcome the difficulties mentioned above, do not avoid the influences of culture. Low scores are just as liable to be due to inexperience in arranging bricks or in interpreting conventional drawings as to having low inborn mental powers. 'Culture-free' tests are probably impossible to devise, anyway. Indeed there is some justification for pressing on with the use of Western-type conventional tests in developing countries like Africa, because African aspirations are in the direction of a western technological civilization.

Last, as implied above, no perfect test of real intelligence has been devised. A test which involves writing answers on paper could favour children from homes with books about and where paper and pen are in everyday use, and handicaps a child from a working-class home where there are often no cultural 'tools'. There is, however, one line of approach which might help us to measure at least some aspects of intelligence objectively. After the brain has received a message through the eyes about, say, a flash of light, it produces electrical waves, whose frequency can be measured, and there may be some link between the frequency of these waves and some forms of intelligence, but, as we have pointed out earlier, intelligence has more than one dimension and does not involve speed alone.

Precise and objective tests, like the one above, of facets of a child's intelligence may well come in time. By that time, perhaps in a century, the basic work on the biology and biochemistry of the brain may well have been done and specific remedies may have been provided for specific defects identified. But at present it must be emphasized that our studies on the inheritance of IQ are crude and so are our notions of intelligence.

THE GENETICS OF INTELLIGENCE

In Chapter 2 we have talked generally about the transmission of high and low ability, and we need to look a little closer now at its genetic basis. Many inherited characters, like the ABO blood group system or eye colour or phenylketonuria, are clear-cut in that you do or do not have an A blood type or phenylketonuria. This is because a blood type in an individual or his eye shade or the disease, if he has it, is determined by a single gene pair. As we stated on page 8, one of the pair might be sufficient in itself to cause the character, overriding the effect of its partner gene; and a double dose of this stronger (dominant) gene would not increase the effect of just one dose. On the other hand a double dose of some (recessive) genes might be necessary to cause the effect, as in O blood group or phenylketonuria.

Such characters as blood groups are called *discontinuous* because, in a group of 1000 people, so many are group A, so many B, so many O, and so many AB. With most human characters there is no such clear-cut variation and they are described as *continuous*. People are not tall or short, stupid or bright, sterile or fertile. These traits and many others are continuously graded. Since intelligence is a graded characteristic, that is a few in a population of 1000 people are very bright, a few dull, but most of us are in that grey zone of mediocrity, it seems clear that variations in intelligence in individuals must be determined by a multiplicity of genes (multifactorial inheritance), not just a pair. Moreover such genes do not seem to have major effects; rather the effect of each gene is small and additive. In other words, each gene makes a separate contribution to the IQ total, but they act together and cumulatively to produce the total score. The graded curve for IQ shown in Figure 4 is not, of course, only due to these genes of small effect but also to the influences of the environment—schooling, background, and so on. Multifactorial inheritance affects variations of IQ within normal limits—all values above the average of 100 and all those below the average down to 70 and probably 45, but as explained earlier other special cases of defective intellect (idiots and imbeciles) may be due to the effects of a double dose of recessive genes.

The curve shown in Figure 4 over-simplifies slightly. There is a slight bulge between IQ 70 and 90 and an excess of IQs at the high end of the scale which kicks out the curve slightly from IQ 130 onwards. Probably the former is caused by the combined effects of poor homes

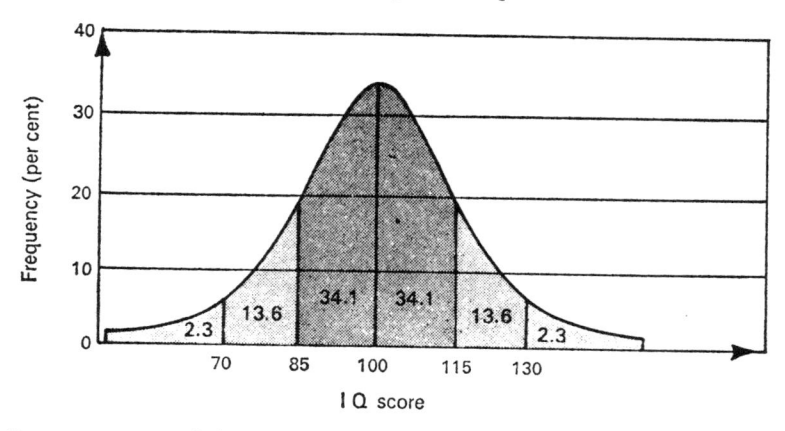

FIGURE 4 Normal distribution of IQ for a population whose mean is 100 is shown by curve. The standard deviation, that is, the usual measure of variation, is about 15 points and the distance in either direction from this mean is measured in multiples of the standard deviation. Thus about 34 per cent have an IQ with a value that lies between 85 and 100, another 34 per cent of the population have an IQ score of 100 to 115 points (*dark colour*). Those with very high or low scores are a smaller part of population: about 2 per cent have an IQ below 70, whereas another 2 per cent have an IQ above 130 (*light colour*). (From Bodmer and Cavalli-Sforza, 1970.)

and of emotional disturbances: the latter by major gene effects that lead to exceptional abilities, just as major gene effects make for sub-normality at the other end of the scale, as we have seen. These irregularities in the curve, may, of course, be due to the characteristics of the tests.

Several other points are worth making in relation to the genetics of intelligence. First, the fact already mentioned that like tends to marry like (positive assortative marriage). Tall people tend to marry tall people; short people short ones; clever people clever ones. In the first choice of a mate the two individuals do their best to size up compatibilities of intelligence and temperament, religion, social standing, and physique. On the whole the dull marry the dull and the gifted the gifted because they like each other's company. Indeed the maintenance of high intelligence in a family like the Darwins or Huxleys is probably due to positive assortative marriage.

Second, if it is true that people of similar IQ tend to marry, how is it that half the children who pass the IQ eleven-plus test for grammar schools are the sons of skilled manual workers and that about a third of the sons of professional men fail the test, despite the fact that the average IQs of the parents of the children are very different. It is because the IQs of children whose parents fall into the extreme groups tend to 'regress' from the extremes towards the mean (IQ 100) of the general population (see Figure 6). Fathers (or mothers) with mean IQs of about 84 or 140 have children with IQs of about 92 or 120 respectively. These values are, of course, averages; some children will be more extreme than their parents but more will be less so. This variation among children, described in more detail in Chapter 6, is, of course, very much the result of the chance reshuffling of the genes for intelligence when gametes are formed, as previously described. This example, incidentally, brings out an evolutionary point: that nature really works towards the 'average' in a population. In an animal population it is the average that counts and deviants are quickly eliminated by selection, but in man, although regression towards the average is an evolutionary anachronism, it still exists.

Our third point is this. Why is it that, if the IQs of children 'regress' to the mean of the population, the shape of the IQ curve of the population as a whole remains the same from generation to generation—and it does? It is because the children from any particular class have a greater spread of intelligence than their parents, so that the shape of the curve for intelligence in the population as a whole remains the same from generation to generation rather than becoming more and more pinched up in the middle. Within this general spread, however, 'class' differences in the genes for intelligence remain. How this happens and is maintained, and what its implications are for society, we must reserve for Chapter 6.

TWINS, ADOPTED CHILDREN, ORPHANS, AND 'GENETIC CLASS'

Studies of twins (and the snags involved) as instruments which help us to understand human individuality have been mentioned in Chapter 2. Some scrupulous investigations by Burt on identical twins brought up separately are worth describing to help us estimate the part played by heredity and environment on the quality of 'intelligence'. Identical twins arise from the same nucleus, so it follows that, barring accidents to the chromosome outfit, or a mutation arising in one egg immediately after it has split in two, each twin must have the same chromosome complement. Yet even though this is so, identical twins, as already indicated, may not be facsimiles because the cytoplasm of the original fertilized egg, when it splits into two, may occasionally not divide quite equally and this might affect temperament and intelligence. Despite these reservations, however, identical twins are more like duplicate human beings than any other pair of people, including brothers and sisters.

Burt studied 148 pairs of identical twins, of which 53 pairs had been split up for one reason or another; the father had died or was in poor health so that the mother was the bread-winner, or the family was too large for the wages coming in. One child of the pair was kept by the real parents and the other was fostered as a baby. 'Overplacing' will be discussed later, but in general it can be said that, adopters being more numerous in the higher social classes, the child is often 'overplaced'; that is, his IQ is often lower than that of his foster parents. Only nine of Burt's separated twins were placed in foster homes of the same social class—unskilled with unskilled and so on—so that the average environmental difference between separated twins is *greater* than it would have been if the foster parents had been picked with a pin.

Despite the differences in environment there was a striking similarity in the separated twins' intelligence as measured by IQ tests. In fact in statistical jargon there was a 'correlation' of 0.87 for intelligence in identical twins brought up apart almost from birth, compared with a coefficient of 0.45 for non-identical twins reared together. (Correlation is a measure of the extent to which things go together. A correlation of 1.0 would mean identity in intelligence, 0 complete lack of identity.) Though the identical twins were separated in different homes, heredity tethered them quite closely with respect to intelligence, and the effect of environment seems to be relatively small. If we went out into a playground of a comprehensive school and picked a few children of the same sex and age, the average difference between their IQs would be about 17 points; with non-identical twins brought up together (no more alike genetically than ordinary brothers and sisters) 12 points, and with identical twins reared apart 6 points. Environment, then, does not seem to make much difference to the general basic intelligence. But according to Burt, environment does affect differences in reading, spelling, and arithmetic, particularly the first two. Coefficients in arithmetic, and reading and spelling taken together, in identical twins brought up separately were 0.70 and 0.60 respectively, but for non-identical twins reared together 0.75 and 0.92. Clearly home influence, how much the child is talked and read to and reads to himself, affects reading and spelling, but arithmetical ability is more determined by native wit.

Criticism has been levelled at this evidence. For instance, why should identical twins, with identical packages of genes, brought up together, differ in 'intelligence' at all, and they do differ (IQ coefficient is about 0.92). This, it is said, tells against the heredity argument. But to be fair, no tests can be perfect and successive IQ tests on the same individual give about the same differences as those between identical twins brought up together. We must remember, too, that identical twins may not always be so identical as once thought.

Other evidence is often quoted too, to demolish the genetical argument; a case of separated identical twins differing by twenty-four points in an IQ test. When looked into, however, this case showed that one twin had had no schooling and the other had been to college. The tests involved were predominantly verbal tests, which do not provide very accurate assessments when schooling and cultural backgrounds are so unlike.

One rather powerful piece of evidence from studies of twins which might well show the effects of the environment on IQ test is the fact that twins average some four to seven points lower in IQ than children born singly, and this is irrespective of home and background. The reduction in IQ could be due to the effects of the environment in the uterus or to the reduced attention parents are able to give to two young children born at the same time. If it is true that conditions in the uterus can cause difference in the IQ of twins and children born singly, then it is not a big jump to make to say that individual differences in the uterus before birth could cause IQ differences in non-twins and might account for much of the total environmental variance in IQ.

Although studies of twins can be used for comparing the effects of differences of heredity and differences of environment, they cannot on the above evidence give us an uncontaminated separation of the two, but do give us an *estimate* of their effects, and we can see that intelligence, when adequately assessed, is largely dependent on heredity.

Adopted and orphanage children can add to our knowledge of the part heredity and environment play in IQ score in two ways. If a group of adopted children is divided into a number of samples in a random way, the average genetic make-up of the children in each sample group should be similar. If the children are placed in different adoptive homes then the effects of the differing environments on IQ score can, in theory, be estimated, just as can be the effect of different quality soils on the response of seeds. All this sounds cold-blooded. Of course no planned human experiment of this sort has been carried out. All the same, such a situation has been found and will be described in a moment.

If the IQ score of adopted children is compared with the IQ of the natural-born children of the adopting parents, this again enables us to judge the hereditary component of the IQ score; each home would give, in theory, a similar environment to both adopted and natural children, and thus if the IQ score of the two differ markedly it should be largely due to the effects of heredity.

The case mentioned of unselective placing is very unusual. The mean IQ scores of the adopted children were found to be very much related to the quality of the adoptive homes, even though the average genetic endowment of each sample of children was supposed to be similar. In the 'good' homes children score on average 112 points, in average homes 105 points, and in poor homes 96 points. By 'poor' is meant not an unfavourable home, but one less favourable than the other two. This suggests that home environment can modify IQ test score by 16 points, not much when we remember that the normal variation which occurs in a population spans at least 100 points: but still large enough to encourage us to create environments for children that will raise an IQ score.

The other observations on the IQ of adopted and natural-born children show that the natural-born children of professional men score on average about 119 points in an IQ test, adopted children in the same home 113; children of fathers who are managers scored 118, their adopted children 112; semi-skilled fathers' 'own' children 101, their adopted children 109. Looking at this human experiment in another way, children reared in foster homes have a very low correlation with their foster parents' intelligence, between 0 and 0.20, even when the adoption agency attempts selective placement, but natural parents and 'own' children's IQ show correlations of about 0.50. It seems reasonable therefore to conclude that the differences between the scores in adopted and in 'own' children are because natural-born children have inherited part of their parents' genetic outfit and therefore resemble them more. As the geneticist C. D. Darlington says, it seems that a social class is stamped on the child by heredity at the moment of conception.

Studies with orphans provide further evidence that our genetic dowry powerfully determines IQ performance and that a genetic class is 'inborn'. One British study shows that the IQ test score of orphans was related to the occupational class of the father. Since, of course, the parents were dead, the average scores of children from various 'occupational layers' had to be used, to compare with the IQ scores of orphans whose fathers, although dead, belonged to a particular known class. In spite of the relatively uniform environment of the orphanage, IQ scores showed a similar *order* to that which might have been expected if the children had been brought up in the homes of their true parents. Thus the orphans (boys) of highly-skilled artisans scored on average 104, orphans of semi-skilled workers 101, unskilled workers 96, and a mixture of orphans of dock labourers, pedlars, and gypsies 98.

Summing up the evidence from studies of twins, and from adopted and orphanage children, on IQ score it seems clear that for a *population* much of the variation in intelligence among school children is due to variation in genetic endowment, and some, perhaps no more than 25 per cent, is also due to variation in quality of school and home. As for *individuals*, the only safe generalizations to be made in this complex field of human behaviour are:

(i) that their genetic outfit endows each with a certain potentiality which may or may not be realized as they grow up; and
(ii) that it sets an upper limit to what each can, in the best circumstances, achieve. As Burt says, 'neither knowledge nor practice, neither interest nor industry will avail to increase it'.

What is important for society is continually to improve the quality of education and home background so that under-achievement due to poor environments is minimal. And of equal importance to us is to try to recognize the needs of individuals and know that it is *just* to treat different people differently so long as each is treated as well as possible. This line of argument is continued in Chapter 8.

4: The Parable of the Talents— the Genius and the Gifted

HEREDITY appears to share out the talents like the master to his servants in the parable; some get five, some two, and an unfortunate few get but one. As in the parable, the talents of the mind can lie buried to tarnish unexposed to books, conversation, and learning, while, on the other hand, no amount of learning will make one talent grow to five. If the five talents of genius are awarded by heredity, how does this happen? Such genius quality (and here this is not taken to mean high cognitive ability only) is due, as we have suggested, to an extremely improbable arrangement of genes inherited from the parents. Genius is a chancy quality because, as we have seen, even the children or the brothers and sisters of geniuses are not often geniuses themselves. The star-quality loses its brightness. This is because when the genius reproduces, his unique gene outfit is taken apart in gamete formation and new ones are built up with the mother's genes, and it is very unlikely that the genes that gave the particular quality will turn up again. Similarly the brothers and sisters of a genius inherit unique gene arrangements from the parents and these are unlikely to give the quality again.

How is this theory borne out by actual examples? Geniuses do seem to turn up out of the blue but, if we examine the records, their parents are intelligent and have character traits which when put together by genetic recombination in an astronomically improbable association produce genius. Bunyan, Nelson, Faraday, Hogarth, Dickens, George Eliot, and many others turned up out of the blue like this.

Bunyan's father was a blacksmith, a member of a village family of long standing that had been going down in the world for some generations and whose mother came from 'decent and worthy' people. Though poor, the parents had a good reputation. 'Post-mortem' IQ estimates assessed by pieces of writing done as a child on Bunyan are 105 as a youth but 120 as a man. This is not exceptional perhaps, but he had the gift of the discriminating eye, the broad sympathy and keen sense of humour which accompany that gift, and his *Pilgrim's Progress* sprang into being 'effortless and fair like a flower'.

Faraday's father, like Bunyan's, was a blacksmith of good character and his mother was uneducated but industrious. As a boy Faraday was a great questioner. He 'could trust a fact and always cross-examined an assertion'. As well as this he was very imaginative. Yet his post-mortem IQ as a boy was only around 105; as a man 150. Both men illustrate the principle of uncertainty described on page 13. The uncertainty here shows how ordinary parents can have extraordinary children by genetic recombination: but both examples also show how limited IQ studies, especially post-mortem ones, are!

Some famous men, however, have not suddenly cropped up. Rather they shine out of a galaxy of other stars not quite so bright. Charles Darwin and John Sebastian Bach did not arise in a single step but in a series of steps; over generation after generation, the new, outstanding gene combination was put together. With the 'short' family trees that gave rise to Shakespeare, Faraday, or Bunyan, the new genius type soon disappears. But with the 'step by step' family tree with its associated assortative marriage, a lineage of famous people arises like the fifty-seven Bach musicians, the thirteen Wyatts who were architects, and the eight Bernouillis who were mathematicians.

All the outstanding men mentioned above must have possessed a reasonable inborn general intelligence besides certain distinct ability factors. But, as expected, traits of character other than intelligence contribute to high achievement. 'Personality' may play its part and often so may sheer *technical skill* (for example in playing an organ or violin or winning an Olympic event). Galton nearly a century ago noted the three qualities, ability, zeal, and the capacity for hard work, as the essentials of genius, and he postulated that they were inherited (IQ + effort = merit). He went on to say that the people in question could no more repress their genius than a spring could be blocked by a stone. They are 'haunted and driven by an incessant craving for intellectual work. If forcibly withdrawn from the path that leads towards eminence, they will force their way back to it as surely as the lover finds his mistress.' Galton made quite clear his awareness of the difference between superior skills acquired through favourable circumstances and achievement chiefly derived from the inner pressure of hereditary endowment. Here, as we have observed before, nature overrides nurture to give the Bunyans, Dickenses, and Faradays.

On a lower level, every schoolmaster can tell of bright, responsive, talented children, marred by lack of stamina. L. M. Terman's studies of gifted children noted that not all intelligent children achieve equal eminence as adults because some have not the necessary drive, persistence, confidence in their ability, and strength of character.

More modern studies than these, on fifteen-year-old boys, lend weight to the importance of personality in achievement. These studies done by L. Hudson, show that about a quarter of future Open Scholars at Oxford and Cambridge fell within the *bottom* 30 per cent of an IQ test sample. These future Open Scholars are distinguished not by their test scores but by their tendency to work hard and by their avid interests outside run-of-the-mill school work. And similar to these rather surprising results is the fact that 23 per cent of Fellows of the Royal Society in university appointments at Cambridge in 1955 passed at only second or third class honours grade in their finals—about the same percentage who took the same subject in the same year went into research but did not receive the scientific accolade of F.R.S. Clearly something is necessary besides mere brain power (though the results of finals may, of course, reflect the nature of the finals examination).

It has been argued by some that anything above a moderately high IQ of 115 to 125 has little bearing on future greatness. A fair amount of brains *plus* character are probably what counts—all these qualities that Galton underlined nearly a century ago—plus imagination and luck.

L. M. Terman's mammoth study of gifted children and what happened to them later in mid-life underlines the importance of temperament and personality in making good or not. Even during their 'gifted' childhood the least successful adults scored low ratings for emotional stability, perseverance, self-confidence, and social adjustment. True to their nature and thus their genetic make-up, they changed their jobs far more frequently than the successful group, found it harder to settle down, their marriage rate was lower, their divorce rate twice as high. The most significant difference between the successful and unsuccessful group in mid-life was a difference in the drive to achieve. And to pile it on to the majority of us who have not got these superior gifts, heredity has generally dealt out the extra talent of good physique to the gifted. Gifted children are no pimply, weedy swots. Rather they tend to be big and strong, to have better sight and hearing than most, stutter less, grow their teeth faster, and mature more rapidly than the majority. Perhaps where behaviour problems arise in school it is due to our fault in not recognizing their ability, and compelling them to mark time while the rest catch up. The result can be boredom or laziness. Worse; the bright boy may hide his light under a bushel, afraid of the other children's twitting him for being teacher's pet.

Perhaps the clearest and most convincing evidence that talent is inherited comes from the fact that ordinary parents can have extraordinary children. This can be explained by genetic recombination or to give it another name, 'hereditary uncertainty'.

Talent, then, can be a product of hereditary determinism, similarity between parents and children, and hereditary uncertainty due to recombination of parental genes. We have implied that character traits as well as intelligence are inherited but as yet have given no hard evidence. Some will be found in Chapters 9, 11, and 12.

5: Men, Women, and IQ

LOOKING back on the geniuses of the world, one cannot help but note the preponderance of men. True, there have been great women novelists, but in the arts of painting, composing, and sculpture, and the fields of science and medicine, great men have been rare, great women rarer. No doubt this apparent shortage of female genius is due in part to the subjugation of women in the past, but perhaps this is not the only reason for their 'averageness'. Not only do men contribute more geniuses to the population, they supply a greater share of defectives, criminals, stutterers, and suicides. They seem to be more extreme than women whatever is being tested. In intelligence tests, young men achieve the highest and the lowest scores, while women tend to gain average scores. Clearly, there must be a biological basis to this 'averageness'.

From the instant of conception woman is different from man, as we said in Chapter 1. Women have two matching X chromosomes but men have only one X, and its partner, the Y, is much smaller. So cytologists looking down the microscope at the forty-six chromosomes of a woman have no difficulty in arranging them in matching pairs. In man, twenty-two of the pairs match but the XY pair is odd. Man, then, is the 'hybrid' sex in respect of the X chromosomes; he has about two-thirds less chromosome material than woman in this pair of chromosomes since the Y, which is virtually an X curtailed, is about a third as long as an X. The woman, on the other hand, is 'true breeding' in respect of the X's. She is evenly balanced in her chromosome outfit. Perhaps this is why men are more deviant in every way than women. They are certainly more variable in shape and in anatomy than females. In his ability to live man is more diverse; his death rate is higher, both before and after he is born. And he is more often born defective. In children, girls appear to be better protected than boys against the effects of malnutrition and illness. They are less easily thrown off their normal growth curves too, perhaps because the two X chromosomes provide better regulatory forces than one X and the small Y. And psychologically it is well known that stress in families affects boys to a greater degree than girls.

All this has evolutionary significance. Biologically the male is only needed to fertilize the female and to protect and provide food for the family until its members can protect themselves. This means that the male is required to respond to highly variable opportunities and stresses. Survival may not be important, but he must be a good fighter, hunter, taker of risks, and altruistic. He must have high-speed muscles and low body fat so it is no wonder that his selected characteristics do not stand the stresses of damage, exposure, and starvation that women can endure.

The female, on the other hand, produces and cares for the baby, and babies need their mothers for the first months and even years of life. Perhaps in the female the greatest stress of natural selection is likely to come on factors favouring survival during the reproductive period, when survival is a matter of two people, mother and foetus. Anything that lowers physiological efficiency will have strong selection against it. Biologically, then, it is an advantage to the woman to be built on average lines as she is more certain to survive (at least in primitive conditions) than a deviant. Female averageness with respect to mental and physical characteristics may be a built-in survival kit, a left-over from the time when the environment was tough, as also is the fact that the greater proportion of human variability resides in the male.

All this does not explain 'masculine' and 'feminine' intelligence if there is such a thing, but the XX–XY relationship may provide the basis for it. As is relevant also to many other genetically-determined

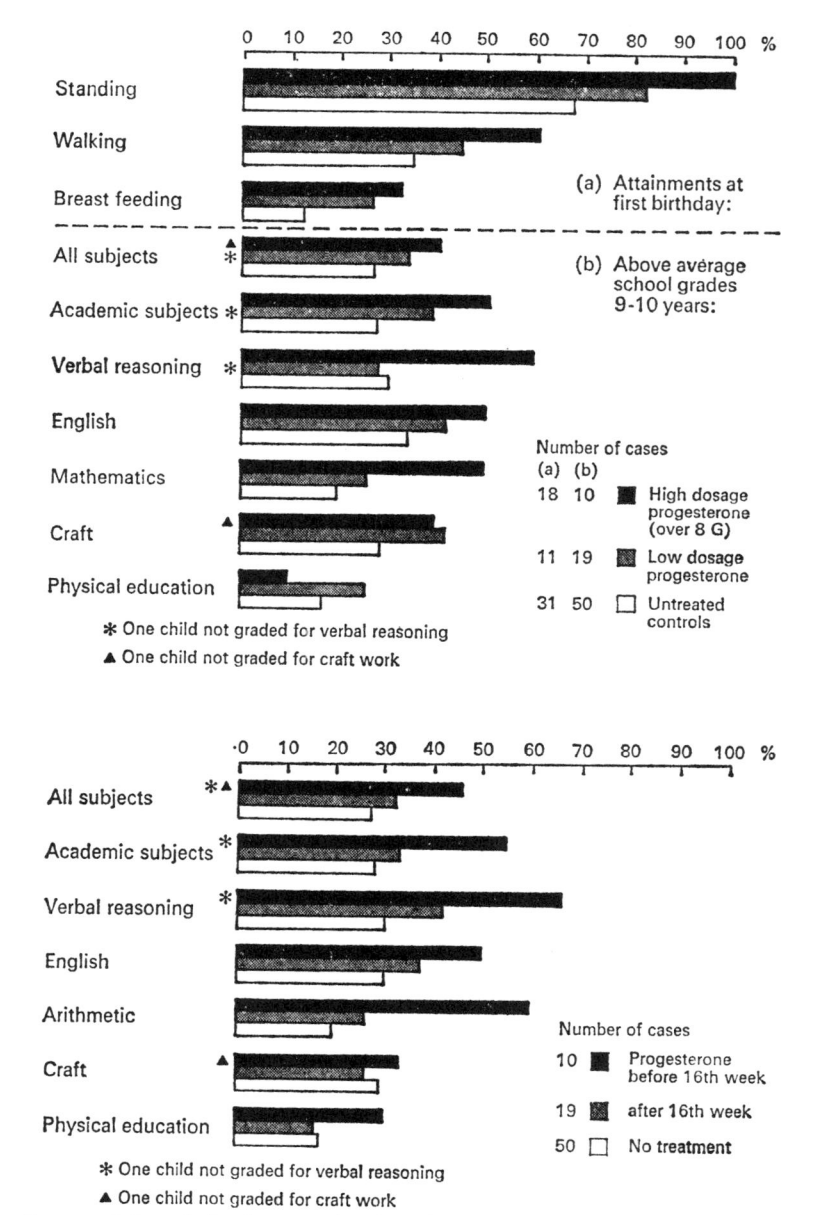

FIGURE 5 A study has been made of 90 children whose mothers received antenatal progesterone, compared with controls. More progesterone children were breast-fed at six months, more were standing and walking at one year, and at the age of 9–10 years the progesterone children received significantly

conditions, man and woman only become man and woman after a long and complex process of growth, development, and maturation.

The agenda of sex development set out by the chromosomes is worked through in two bursts. At first the baby in the womb is neuter but, by the end of twelve weeks, if it is XX, the girls' external vulva, and if XY, the boys' penis, have been moulded out of similar tissue. In this time too the internal reproductive organs have been formed: ovaries, vagina, uterus, oviducts in the female; testes, scrotum, sperm ducts in the male. Ovaries already contain some primitive eggs, and testes some sperm cells. The second phase, which starts in adolescence, depends on chemical hormones produced by the ovaries and testes, apart from eggs and sperms. The hormones, about which we shall say more in Chapters 18 and 19, may vary in quantity and quality from individual to individual. They are not dependent solely on the simple XX–XY switch but also on variation in *all* forty-six chromosomes. It is this variation which prevents sharp male–female differences and gives the continuum between the sexes in secondary sexual characters (breasts, fat distribution, width of pelvis, depth of voice, distribution of hair) and temperament and probably intelligence.

What has all this to do with masculine and feminine intelligence? All the chromosomes except the Y, which is only present in man, are common property to male and female in that they are shuffled about, in and out of male and female from one generation to another. At any one moment in time, for instance, two-thirds of the X chromosome pool is in the female, a third in the male. If, then, there is a difference in intelligence, it must stem basically from the difference in the sex chromosome outfits, but as we have indicated, sharp differences are prevented by the fact that forty-five chromosomes are common property to both sexes.

The fact that the sex chromosomes have some direct effect on 'intelligence' is shown by an abnormality caused by the loss of an X or

better gradings than controls in academic subjects, verbal reasoning, English, arithmetic, craftwork but showed only average gradings in physical education. (Top chart.) The development and intellectual advantages were all related to the dose of progesterone received by the mothers, those receiving over 8 g being related to early walking, standing and better school gradings. The intellectual advantage was greatest in children whose mothers received progesterone before the 16th week of pregnancy. (Bottom chart.)

From Dalton, K. (1968) 'Ante-natal progesterone and intelligence'. *Brit. J. Psychiat.* 114.

a Y chromosome during gamete formation, leading after fertilization to a sterile, sexually abnormal, and sometimes mentally defective, individual with a chromosome outfit of 44 autosomes + XO = 45 chromosomes in all. The absence of a Y chromosome, of course, determines that an individual is female. Some of these unfortunate people, while not mentally defective, are very poor at tests involving shapes and patterns, and this is sometimes so severe as to be popularly known as 'space-form blindness'. These rare people also report unusual difficulty with arithmetic and mathematics in school despite otherwise normal or superior intelligence. So here is a genetic fault caused by sex chromosome anomaly and imbalance, clearly seen under the microscope, which has quite specific consequences on the cognitive processes. In other words, there is good reason to believe that the sex differences in visuo-spacial ability are under the control of the sex chromosomes. Indeed a male on *average* has a higher visuo-spacial ability than a woman (J. Gray).

There is some evidence that masculinizing hormones (androgens) circulating in the blood and affecting the brain of the baby while still in the uterus, and released in excess from its adrenal glands (due to a genetic defect), are often associated with a higher than average IQ. Recent work by J. Money on the IQs of seventy people with the rare hereditary defect of the adrenal gland showed that only 27 per cent had IQs below 100 (50 per cent of the normal population fall below IQ 100). Although, on average, only about one person in ten has an IQ over 120, almost one in three of the seventy subjects were above that level. Money has also discovered that ten children whose mothers had been given progestin* (to prevent miscarriage) all had above average IQ and six had exceptional IQs of above 130. Progestin is chemically very similar to androgen and some of it inevitably found its way into the bloodstreams of the unborn children. Research has thus revealed a correlation between high levels of androgens or progestinic substances present in the developing infant while still a foetus in the uterus and high IQ levels subsequently manifested by that individual. (See Figure 5.)

Genes on the Y chromosome, as we have indicated, cause the development of the testis, and at about the twelfth week of development the testis produces androgenic hormone which in turn causes the penis and scrotum to form. Could this hormone and others affect the developing brain of the normal child (in perhaps a similar way to

* The synthetic equivalent of progesterone, the pregnancy hormone.

the abnormality mentioned above) to cause in normal development a physical difference in the brain of men and women? (See pages 169 and 178.) Sharp differences would be ruled out because of the buffering effect of the common chromosome pool. Hormones certainly do have an effect on learning capacity in rats. In the absence of thyroid hormones, the nerve cells of the brain in immature rats do not develop good inter-connections and this may account for poor learning capacity.

But what evidence is there that a physical difference between the brains of men and women exists? Weight of brain in a female is only slightly less than in a male, and this is only relative to her smaller bulk. We can approach the problem of what is essentially the physical basis of intelligence in a round-about way. Sherrington declared that in any given individual nervous tissue, like all other tissues—skin, hair, muscle, and bone —tends to be of the same quality throughout with minor variations.

Microscopical studies of the brains of mental defectives and normal people show the nerve cells of the defectives' brains to be stunted and feebly branched and 'deficient in number and irregular in position' (the efficiency of man's intelligence is thought to depend on the number of cells in his cortex, or grey matter, and the number of inter-connections between the branches of the nerve cells). Here then is a physical link between low IQ and brain structure. Taking this a step further into the realms of normality, recent studies emphasize the enormous variation in the detailed cell structure that exists between the brains of different people. This finger-print individuality of the brain is something we must return to in Part 2. The variation in cell structure in the brains of the defectives mentioned above seems, then, to be only one end of a spectrum of variation between brain and brain, and all this indicates a structural origin for differences in mental endowment.

No evidence exists at present which points to any physical difference between the brains of men and women. Yet it is absurd to think of the sexes as being equal. They are not. Although in almost every character they overlap, they are *different*. As C. D. Darlington writes, 'each has its value, a value best appreciated by the other, a value the result of long selection and adaptation'. The education of girls, as J. Newsom pointed out, is based on the (false?) assumption that women have a similar kind of intelligence to that which is needed and possessed by men. As for other characters, selection in the past must have worked in different ways on intelligence and behaviour for the woman who had the babies, and for the hunting and protecting man who only begot the babies! If there is weight to these evolutionary arguments, twentieth-century girls, it seems, are too often prepared for a life most of them will never lead or wish to lead and which 'it would be disastrous to the community if they did follow' (Darlington).

While there is no scrap of evidence for a physical difference between the brains of men and women, there is evidence of their differing attitudes. First, Burt's inquiries with gifted London school children showed that boys had marked preferences in reading, for science, engineering, aeronautics, and the latest inventions, and as lighter reading, travel, detective stories, and tales of the sea. The girls liked poetry, biography, history, and natural history. Boys' hobbies were building complicated meccano models, radio sets, toy aeroplanes, telescopes, playing with chemistry sets and electric motors, and collecting stamps and fossils. Girls liked painting, drawing, photography, and writing poetry, but all this can of course be explained by tradition in rearing.

Next, there is the evidence of intelligence test results and class of university degree. Women tend to do better in verbally biased items in tests, and men in numerical and diagrammatic items and also in problems that demand an understanding of mechanical principles. Men clearly do better in occupations which require a high level of visuospacial ability.

In the University finals at Cambridge between 1920 and 1952 women display their 'averageness'. They gained the highest frequency of Thirds and Seconds in mathematics and men the highest frequency of Firsts. In the natural sciences men were awarded the highest percentage of Firsts and Thirds; women again earned the largest proportion of Seconds. Even in English, despite the fact that women score better in verbal tests men gained more Firsts *and* more Thirds than the women, who scored a disproportionate number of Seconds.

The male-science/female-arts split (seen in many of our schools, particularly in co-educational schools) may, as A. Heim suggests, be a congenital difference rather than a social artifact, but it is hard to tell as long as society continues to treat women as intellectually inferior to men. On the arts/science bias the facts are that out of 62 519 undergraduates admitted to university in 1970, 19 909 were girls: 17 100 of these are studying arts, sociology, commerce, and allied subjects, and out of 10 217 students admitted to engineering and technology courses only 243 were girls.

Is all this male–female difference really due to our culture? There is a strong possibility that it may be so. In Russia where women have more equal opportunities and where there are nurseries for children, about a third of the engineers and lawyers and two-thirds of the teachers and lecturers are women. There are more women doctors in Russia. Girls in western societies are given dolls and dish-washers rather than meccanos and model aeroplanes, and are encouraged to be passive, responsive, and eager to care for children. Boys, on the other hand, are brought up with the idea that their goal in life is a successful, stimulating career. Perhaps all this underlines one thing, recognized by Byron; that 'Man's love is of man's life a thing apart, 'Tis woman's whole existence'. And at the basis of this lie the brain and hormone system, both genetically determined: the brain discriminating, choosing, recognizing a mate whom we could get on and work with and whom we love (and this is uppermost in a woman's choice), and the sexual hormone system which in a man drives him towards a woman. In the woman the effect of the female hormones in creating desire is more dependent on the availability of a man whom she can love and who seems worthy to her.

When applied to individuals these 'laws' sometimes break down because they only apply to the average part of the curve for genetic variation. Some women do seem to develop a rather masculine kind of intelligence and temperament and some men are the reverse of what is 'expected'. But this is part of the range of variation which makes sexuality lie on a continuum. This is created not only by the XX-XY chromosomes but all the other chromosomes varying and playing their part in delineating the whole character of a man or a woman. But a sound general principle could be applied to the intelligence of the sexes with advantage: *equal, but possibly different*. This is not an insult smacking of apartheid but a statement based on our available knowledge of genetics, physiology, and psychology.

Third, what do women excel at that men don't? Women in our society seem to be good at picking up tiny flaws and make good industrial inspectors. They seem to lack the fundamental ability to organize large units without worrying what is going on at floor level. But to be fair, the only way women might be able to progress in industry is to take up some hitherto neglected speciality. As we all know women make first-rate nurses, infant teachers, social workers, and telephone operators. So, with the exception of industrial inspecting, women thrive in occupations which require an interest and a liking for people. All these jobs deal with the concrete rather than the abstract, and all, in their way, need meticulous attention to detail. But let Samuel Johnson have the last word. When asked which are more intelligent, men or women, he replied, 'Which man, Sir, which woman?'

6: Class and Brains

IS a social class stamped on a child at the moment of conception? In the past there were very strict religious and class barriers to marriage, so that the different classes knew their place in society. Even with brains and drive Thomas Hardy's Jude could not get into Christminster to become a scholar, because he was a stonemason. Perhaps the stronger class prejudices of a hundred years ago or more kept the classes more distinct genetically than they are now when there is more mobility between them. Of course there has always been some movement between classes; up for the exceptionally able, down for the exceptionally stupid. Since the last war changes in the structure of the educational ladder in particular, and a less rigid attitude about people's family origins, have shifted the emphasis from class 'inbreeding' to much more 'outbreeding'.

But to return to our first sentence. Despite the fact of a recently-accentuated shake-up of genes in the classes by outbreeding, there is still a fair amount of genetic difference between the various social classes. (The definition of classes is arrived at purely by an assessment of economic and professional standards.) Middle class and lower class differ by about fifteen IQ points in verbal reasoning, number and space tests. But the lower class, disadvantaged child seems quicker on the uptake at picking up the names of children in his class and at learning playground games than a middle class child of a similar low IQ (60–80). Perhaps nature gives, on average, the unskilled man's son a greater facility for the concrete, and, considering the job opportunities that modern western societies will offer him fifteen years later, who can say that nature is unwise? Indeed, although intelligence is vital for the survival of a race or society, it is not necessary for all members of the society to be very clever because a complex society has niches for the bright and the dull. The less able may benefit from association with talented leaders. Selection would prevent a race of idiots evolving but it would not eliminate them from a society which has a need for unskilled workers.

This genetic stamp of 'brightness' or 'dullness' or simply average 'grey' in individual adults does not always indicate that the owner of the brains himself came from the class whose mean IQ he is nearest to, even though he himself may now belong to that class. To explain this we need to look at some recent work by Burt on class differences and IQ in London school children and their parents.

On an IQ test with a mean score of 100, the mean IQ for higher professional fathers (university lecturers, top doctors, lawyers, educationists, and so on) is around 140; lower professionals 130; clerical workers 116; skilled workers 108; semi-skilled workers 98; and unskilled workers, employed in coarse manual work, 85. (See Figure 6.) These bald facts, though, hide the variation beneath them. Thus, although the IQ score of a top lawyer is probably around 140 since he belongs to social class I, it could be higher or lower. The point is that each social class contains individuals who cover quite a high range of intellectual ability. In technical language, the 'standard deviation', which is the usual measure of variation about the mean of the IQs within any of the six social classes of adults mentioned above, is 8.6, almost three-fifths of the standard deviation for the entire group, which is 15. So a predisposition towards a social class is on average printed on a child at the moment of conception, but the child of a skilled worker might easily have the genetic potential of the child of a university professor, for the children of the adults whose mean IQs are given above have a spread of IQ almost as great as of the general population: 13.2 compared with 15. Indeed in a population only one-quarter of all very bright sons are born to very bright fathers, half are born to moderately bright fathers, and one-quarter are the sons of average fathers. If we look at Figure 6, this shows that the higher professionals (mean IQ 140) have children whose mean IQ is around 121; clerical workers (mean IQ 116) have children with mean IQ of around 108; but unskilled workers (mean IQ 85) have children on average brighter than the parents with a mean IQ of 93. Such are the gifts of heredity; golden parents may have leaden children and vice versa.

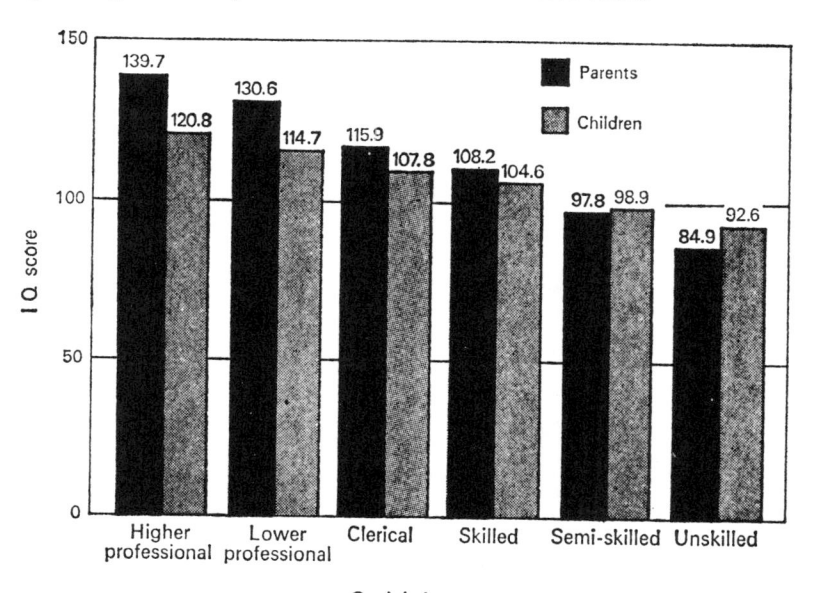

FIGURE 6 Social class and intelligence are closely related, as a study by Sir Cyril Burt of the University of London indicates. Above the mean of 100, children's IQ tends to be lower than that of parents. Below it children's IQ tends to be higher. Social mobility maintains distribution because those individuals with high IQs tend to rise whereas those with low IQs tend to fall. (From Bodmer and Cavalli-Sforza, 1970.)

Why then do the classes maintain their 'steady state' with respect to IQ from one generation to another? Because of social mobility, or better, social promotion. Individuals with high IQs and strength of character tend to rise in society, those with low IQs and negative character traits tend to fall. Burt reckoned that to maintain the existing IQ level in the various classes, 22 per cent of the children of any one generation would have to change class, a figure well below the observed inter-generation class movement in Britain, which is about 30 per cent.

Individuals, then, take their brains, characters, and their breeding potential from one class to another, but the children of these individuals will not necessarily all stay in the same class as their parents; some will, but others will rise or fall.

It is a good thing that the chromosome machinery spills out different talents and IQs in abundance. If it did not, the so-called 'working class' would not have the vitality it does but would become a self-perpetuating caste of men with spades, picks, and drills, while the rest would form their own self-propagating groups. Without considerable variation in intellectual ability and character (which is at least as important as intellect in social promotion), the mobility that we see going on could soon exhaust a class of its potential, and the behavioural defects of each inbred social class would, by the operation of the different upbringing of each, be constantly exaggerated.

Take the 'working classes'. The 11-plus, or the 'scholarship' examination before the 1939-45 war, gave many a clever, working-class lad a golden opportunity at grammar school. They still well up as strongly as ever from the manual, semi-skilled and skilled workers. The trouble is that these clever lads stay in the intellectual élite where they often end up, and who can blame them? But in doing so they deprive the humbler classes of the leaders they badly need. Consider Ernie Bevin, a great statesman and trade union leader. If Bevin, as Burt rightly says, had got a scholarship to Rugby and then Oxford and passed for the Higher Civil Service like so many of the ablest graduates of his day, 'what a loss that would have meant both for the trade union movement and for the nation as a whole'.

We have seen then, in this chapter, that the higher the occupational class, the higher the average measured IQ and thus the reservoirs of genes for brightness and dullness lie at different but associated levels in the population. And these levels are maintained by the flow of brains and breeding potential from one class to another. Intelligence, as we have stressed earlier, does not depend entirely on the genetic outfit of the person tested, but on the opportunities he has had, in particular the attention, interest, love, and intellectual stimulation (or not) of his parents. And to stress these points in relation to class, working-class mothers, whose environment tends to press harder on them, do not smile with their infants or play with them, or reward each step towards maturity with approval as much as middle-class parents do. Nor do they talk to them as much or care as much about school progress, and all these factors can be important in depressing intelligence. Yet in the working classes, the child victimized by the break-up of a marriage is more easily taken into the bosom of a real family and given a rôle to play—all important in the development of IQ—than in other social classes.

Perhaps the pre-school child who sits on his father's knee before he goes off to the office at 8.00 in the morning while he reads and discusses the newspaper, will develop a better capacity for abstract conceptualization than the pre-school child whose father rushes out to work at 7.00 in the morning to join the other labourers at the factory site. But to be fair, in an affluent society, many children of 'rich' parents are more often culturally and socially deprived by broken marriages than the working class child where the 'extended' family responsibility takes him in, as indicated above; to be poor is not necessarily correlated with lack of stimulus from books and conversation.

Perhaps, too, a culturally rich infancy, whatever social class, may have some permanent effect on the functional qualities of the nervous system (see Chapter 11). The hard fact is, however, that by the time a middle-class child reaches school he may be 25 IQ points on average above the IQ level of a working class child. Despite cultural factors we can probably say that genetic factors are responsible in Western Europe and America for nearly three-quarters of all the individual differences in intelligence and about 25 per cent of these differences are environmental in origin; in other words heredity is three times as important as environment. And this to some extent shows in the class a person occupies.

7: The Intelligence Quotients of Black and White

SOME rather painful statistics on the IQ gap between white and Negro communities in the United States of America recently have helped to prime a major explosion on race relations in the States and, of course, its repercussions have been felt elsewhere as all people are more sensitive about their brain power than their looks. At the end of the last chapter, it was emphasized that in Western Europe and America perhaps 75 per cent of the variability in human intelligence derives from heredity. This is not a universal figure. It is not necessarily the same for Indians in India or Chinese in Hong Kong. This is why we have been careful here to state that the work on IQ has been done in America on whites and Negroes. But to return to the IQ gap. The massive compensatory education programme for socially deprived children in the United States of America, many of whom are Negroes, has been declared almost a failure because of the acceptance of the hypothesis that there is a substantial genetic component in the IQ differences between black and white which no amount of pre-school and later education can alter. Arthur Jensen, an eminent American psychologist, compiled the evidence with scholarly reserve.

Much of the evidence he cites for the heritability of intelligence in individuals shows, as do the preceding sections of this book, that environmental differences account for considerably less than half the variability; that is to say that the genes account for more. Jensen believes that only extreme deprivation (and we must go into this later) can keep a child from performing up to his genetic potential, but an 'enriched' programme of education cannot push the child above that potential. What do the raw statistics say about the IQ gap? They show that the IQ of Negro populations is about 15 IQ points below the average of the white population but 15 per cent of Negroes exceed the white average. It is important to stress the difference here between populations and individuals. As far as we know the full range of human talents is represented in all the major races of man and at all social levels, and clearly it is unjust to allow skin colour or class to affect an individual's place in society and treatment of him by society. Race differences in intelligence come up, not when we are considering individuals (genius is neither lacking among Negroes nor universal among whites) but when identifiable groups or populations are brought into comparison with one another.

Comparisons like these with populations bring out any average differences and their magnitude, and as a result specific questions can be asked about such differences. What is the significance of the difference, medically, socially, educationally, or from any other standpoint that may be relevant to the characteristic in question? Such questions are asked because there is social demand for programmes that concentrate on compensation for group handicap.

The fifteen points average IQ difference between black and white populations could be due to gross deprivation as well as to genetic differences and is certainly in part due to both, but the difference seems to be a permanent one and is true not just for one IQ test but over eighty-one different tests of intellectual ability. A more extreme average difference between IQ of Negro and white populations is based on a vast test programme given to 18 000 black children in elementary schools in Florida, Georgia, Alabama, Tennessee, and South Carolina. Compared with a representative sample of whites of mean IQ 102, the IQ of the black sample was 81, a difference of 21 IQ points between the two population means, but there was considerable overlap between the IQs of the two populations.

The preceding section of this book has shown that genetic differences powerfully influence IQ in individuals. Is it a reasonable idea that the races of mankind, or better the *sub-species** of the human species, differ in intelligence? Most biologists would agree that humans, like other animals, if geographically or socially separated into different populations over hundreds of generations, are practically certain to differ slightly in their gene outfits. Mutations occurring at random in the different populations, if useful, will spread rapidly through them. And mutations incorporated in the gene combinations of the different sub-species of man are probably adaptive in some way—to climate, altitude, or disease. These adaptive genes, however, are embedded in gene outfits which give all human beings a general flexibility which allows them to live in a wide range of environments. So, although the human species is one, in that its members, irrespective of skin colour, can interbreed, a pile-up of different genes here and there gives this or that racial feature. All the genes and chromosomes, however, are common property to the species, and intermarriage following migration and conquest has prevented the different sub-species of man from becoming different species, that is, being unable to produce fertile offspring.

We know that there are genetic differences between the sub-species in skin colour and in the frequency of the A, B, and O blood genes and genes controlling other blood group systems. These relatively simple polymorphisms (a polymorphic gene is one of a group that accounts for variability in a particular characteristic) are very useful in understanding the nature and size of the differences or similarities among human sub-species, but the inheritance of the more obvious face and body traits is complex and not really well understood genetically, although clearly they must have a substantial genetic component. On this argument there is no reason to suppose that the brain and behaviour of the different sub-species should be exempt from the effects of natural selection, which is the principal agent of genetic change, and that the many genes controlling intelligence might have different average frequencies in the different sub-species.

It is likely, however, that the races, even though isolated from one another during evolution, changed in parallel as far as the evolution of the brain and intelligence are concerned.† If not, why the considerable overlap in IQ scores? Primitive man had to have intelligence to invent and use tools, weapons, and shelters, and as each tool or weapon was invented it sharpened the selection against those who could not

* For a discussion of 'race' and 'sub-species' see the companion volume, *Biology and the Social Crisis*.

† See *A Natural History of Man* (London: Heinemann Educational Books, 1969.)

exploit it. Thus the very use of such implements must have selected individuals who had the brain power to use them. In other words each invention gave greater advantage to more ability of a similar kind. This self-exaggerating effect with a positive feed-back on the brain and intelligence must have led to gradual evolution in one direction. No race or population could escape this mechanism of evolution of brain and intelligence, on which its survival depended. Indeed it is difficult to think of a race in which intelligence, co-operation, and physical health would not have positive selective value. Hence it is very likely that natural selection would oppose the establishment of major heritable behaviour differences between races. Why then is there an IQ gap between black and white population averages? The following points are worth bearing in mind when considering the question whether genetic factors play as large a rôle in the difference between races or groups as we have seen they do within them.

1. 'Comparable' groups such as Negro and white in America have never been equal (and probably never can be) for simple physical health, life expectancy, and nutrition during pregnancy and very early childhood (one year and under); these factors may well have a drastic effect on IQ. Observations on female rats fed on protein-deficient diets before and during pregnancy show a considerable reduction in total brain DNA content of the offspring and hence presumably a reduction in the number of brain cells. The reductions are linked with behavioural deficiencies. In man, if comparison is viable, lowered nutrition could be a reason for substantial IQ reduction. More direct evidence comes from the IQ of severely malnourished children and its link with brain growth. Brain growth is largely a process of protein synthesis and its protein requirements in the womb and the first year of life are avid for it grows at the rate of one to two milligrammes per minute. Indeed, in the first year of life 70 per cent of the maximum adult weight of the brain is attained. Extremely malnourished South African coloured children were some twenty IQ points lower in IQ than children of similar parents who had been well fed. In New York City, women of low socio-economic status were given vitamin and mineral supplements during pregnancy. These women produced children who at four were on average eight points higher in IQ than a control group of children whose mothers had been given placebos during pregnancy.

If poor feeding does take a toll it probably happens in the uterus and early in life, at one year or under, and its effects may be irreversible. The point is clear: dietary deficiencies in pregnancy and very early childhood may starve the brain and permanently lower the IQ. Since the poor socio-economic condition of the blacks is linked with dietary deficiency it follows that the uterus, as suggested earlier, might be decisive as an environmental factor influencing IQ.

2. J. Lederberg, the distinguished geneticist, has made two other relevant points. First, that an astonishing number of children from old slums still turn up with brain damage by lead poisoning from eating flakes of old paint, perhaps aggravated by lead in exhaust fumes. There is firm evidence that elevated blood-lead levels can lead to mental retardation through irreversible brain damage. One in six Manchester children had levels of blood-lead over 50 microgrammes per 100 cubic centimetres of blood—30 microgrammes is the acceptable level for children, 80 microgrammes the danger level in adults.

Second, that some specific genes are related to diseases known to be more prevalent among Negroes. Sickle-cell trait in Africa is a genetic defence against death from malaria because it produces a blood pigment undigestible to the malarial parasite. One dose of the gene confers this protection but a double dose gives the much rarer disease, sickle-cell anaemia. About 9 per cent of American Negroes carry the gene which gives the trait. These carriers of the single dose of the gene are not ill but, as J. Lederberg says, we do not know what the subtler effects of the gene are on the person possessing it, especially when he is under stress. We do not know whether carrier children are more or less intelligent than their normal brothers and sisters. When we have more studies like this we shall be able to claim to have made some tangible headway on the genetics of intelligence.

3. Are IQ tests which have been devised largely by white testers, and administered by them, fair to blacks? On the second point it has been reported that the IQ of blacks tested by blacks is two to three points higher than when they are tested by whites. And on the first point, as we have hinted earlier, tests stripped of culture (such as acquired knowledge, a good command of words or acquired skill at jigsaw puzzles) are hard to devise. A number of studies have

shown that eleven-year-old boys (Canadians, Indians, Eskimos, British, and Hebrideans) score differently in tests involving the ability to grasp patterns and apply reasoning skills, depending on their upbringing—whether it was restrictive or permissive, and whether it encouraged or discouraged resourcefulness. As far as American Negroes are concerned, 'culture-free' tests tend to give them slightly lower IQ scores than more conventional tests. They do better at verbal tests but worse on tests that tap abstract qualities. This may link with genetic factors. If, for the sake of argument, there are different genes that condition how easily a child can learn pictograms on the one hand, or alphabetic syllables on the other (by inborn patterns of nerve cells in the brain keyed to respond to basic stimuli), then it will be quite important for the actual intelligence of a particular child whether he happens to be born in Japan or Britain. Testing of the 'pictogram child' in Britain or the United States of America would not really discover his inborn capacity and his IQ score would be low. Are we discovering the 'real' IQ of Negroes by the application of Western tests?

4. On this point evidence is beginning to mount that different patterns of ability exist across different racial groups. The patterns of scores in spacial tests, verbal reasoning, and number tests, are distinctively different for Chinese, Jewish, Negro, and Puerto Rican children regardless of their social class. Jewish children scored higher in the verbal kind of IQ test, but scored less well in tests dealing with spacial and visual patterns. Chinese and Puerto Rican children did relatively well in the tests for spacial perception but scored low on verbal tests. Negro children fared well in verbal tests but rather poorly in number tests. To add to this evidence of differing racial abilities, when full-blooded Australian Aboriginal children are given a variety of 'conservation' tests of quantity, weight, volume, number, and area, the majority do not pass them even by the time they reach adolescence, but the majority of European children pass them by seven. However, and this is the effective point, the tests are passed by a significantly larger number of Aboriginal children who have one European grandparent or great-grandparent. Do the genes inherited from the white ancestor pass on the capacity to do these tests, and do the genes of the races of man differ to give differing abilities?

5. On the point of the Negro scoring badly in abstract tests of IQ, there is some evidence that when the brain is exposed to deleterious influences, the ability for abstract thought will be the first to go. As the brain convalesces the capacity returns but slowly. Clearly very deprived environments would act against abstract conceptualization. Anybody, Negro or other, having to grapple with the concrete situations of getting enough to eat, keeping a job, worrying about the bills, is forced into concrete thinking rather than abstract.

6. Finally, there is the obvious point that is constantly overlooked: that blacks are discriminated against. At present we have not the means to measure the influence of having a black skin on, say, reading skills or number ability, but a poor self-image because of a black skin might affect ability profoundly. Neither do we know how to cancel the influence, nor even whether we should try. Probably a great many blacks resent taking part in tests which they see as part of the white man's social system, which is broadly anti-black. Such resentment could well depress average IQ scores; indeed possessing the genes for a black skin may, as J. Lederberg says, lead a student with the highest intellectual potential to turn his back on the hard work of learning science and mathematics (which will measure out as intelligence by middle class standards) in favour of black studies that he hopes may meet his more urgent needs in other spheres.

The balance of evidence does not permit us to say that American Negroes are genetically inferior to white Americans in intelligence. Evidence is certainly mounting which suggests they (and indeed a number of racial groups) may have different inherited patterns of ability. But the raw statistics of the IQ gap might be the result of many factors: genes, not only those for intelligence, but for sickle-cell anaemia and indeed for simply producing a black skin with all the bitterness of racial alienation that goes with it; poor self-image as a result of a black skin; low level of teacher-expectation on school performance—low expectancy breeds low standards; poor food, particularly in pregnancy and early childhoood when the growth of the brain is at its maximum; and lack of a stimulating environment in early childhood, for there are data that demonstrate the detrimental effects on the brain of severe early sensory deprivation (see page 125).

In theory the experiment that would settle the questions of the heredity-environment interaction on racial IQ would be, for example, to bring up a large number of Negro children from birth in white homes and vice versa. But even this would not work. Each child would bring into the home its own special environment created by having a black or white skin. No matter how equally the child was treated in the home, society at large would treat him differently.

What is necessary is to improve the lot of the black economically, environmentally, and educationally, and indeed the lot of all the deprived peoples of the earth. We can then get on with determining individual qualities and providing equal and maximum opportunities for their development independent of skin colour. In all considerations of population or group comparisons we are apt to let our minds dwell too much on the difference in the average IQ between populations rather than concentrating on individual variations in educability.

8: Genetics and Education

LORD Butler once said in a television interview in 1966; 'I think that's a very important rule for people in life—to live according to their own nature'. This is one of the points we have underlined about intelligence and human nature in the past few chapters. Here are some points to keep in mind when reading this chapter, and which consolidate much of the argument in previous chapters.

1. That there is no such single quality as human intelligence but there are as many intelligences and natures as there are men. This is because all people, apart from identical twins, are born with unique gene outfits (genotypes).
2. That in almost any character we care to study, like 'general' intelligence or a special aptitude, the genotype sets an upper limit to which an individual can develop.
3. That genetic uniqueness and unique environment interact to produce the unique character—intelligence or height—we actually observe (the phenotype).
4. That we are ignorant as yet about what is or is not the necessary or favourable environment for a particular genotype. Neither do we know which is the necessary or favourable genotype for a particular environment. We know from hard experience that some children in a class need pushing, others need freedom. And

relevant to our next chapter on criminality, poverty may make for larceny in one disposed that way; in another it may sharpen his wits and prepare him for a useful life; and in another an unpleasant environment may enhance genius.
5. That estimates of the heritability of intelligence as measured by IQ tests are valid only for the population from which they have been obtained and are not valid for future populations or new environmental conditions. In short, heritability only measures the relation between existing genetic and existing environmental variation in a population.
6. That although there may well be average differences between races, classes, or groups in a particular character like general intelligence, we ignore the uniqueness of the individual to our great loss.

J. M. Thoday, in two penetrating papers, discussed the importance of genotype–environment interaction and emphasized the point made in 1, 2, 4, and 5 above in relation to IQ, environment and education. He proposed two imaginary situations. In one he considered a genetically simple population of five equally frequent IQ genotypes set in a situation where there are but five kinds of environment, all likewise equally frequent. Without genotype–environment interaction it is clear from Table 1 that if we wished to maximize IQ, all genotypes should have environment E. If the motives were 'equal chances for everyone' then we should attempt to give genotype A environment E; E genotype environment A; B genotype D environment; C, C and so on. If for economic or other reasons we could afford to give the environment that gave the best IQ performance only to a few, we should have to devise selection tests to pick out E genotypes and put them into an E environment. Clearly real situations are vastly more complex, and as we know heredity does not merely 'respond' to an environment like a plant thrusting to the light.

Table 2, although simplified again, is more like the truth and shows a similar arrangement to that described in Table 1 but in this case we have genotype–environment interaction. The IQ phenotypes of one genotype A, vary according to the environment. So here a single genotype produces different IQ phenotypes according to the environment (note point 4 above). And each genotype requires a different environment if it is to develop the highest IQ it is capable of— A requires environment B, B requires E, CC, DD and so on. Here, then,

TABLE 1

		A	B	C	D	E
				GENOTYPES		
ENVIRONMENT	A	80	85	90	95	100
	B	85	90	95	100	105
	C	90	95	100	105	110
	D	95	100	105	110	115
	E	100	105	110	115	120

TABLE 2

		A	B	C	D	E
				GENOTYPES		
ENVIRONMENT	A	80	90	100	110	120
	B	120	110	80	90	100
	C	90	80	120	100	110
	D	110	100	90	120	80
	E	100	120	110	80	90

From Thoday, J. M. (1965), 'Geneticism and Environmentalism' (see Bibliography).

we see unique genotype interacting with unique environment to produce a unique IQ phenotype. Thoday rubs in the lesson with these theoretical examples that no environment is 'better' than any other, and no genotype 'better' than any other. Each genotype requires a different environment if it is to develop the highest IQ of which it is capable.

Relevant to all this are remarks in A. R. Jensen's paper on a big weakness in educational programmes for deprived children in the States. This is that such programmes may be doomed to failure if their success is based on how well these children approximate to middle-class children. At best, he suggests, they are different and they can never be anything but second-rate if they are thought of as potential middle-class children. What seems to be missing in the programmes is an appreciation of genotype–environment interaction as emphasized above. Apparently these children are being given the wrong educational programme to maximize their IQs.

The teaching methods used in compensatory educational programmes, Jensen says, depend very much on whether a child is average or above average in intelligence as measured by conventional IQ tests. It is important to discover and devise teaching methods that capitalize on existing abilities (the ability to learn basic skills, for example, as measured by direct-learning tests) for the acquisition of 'those basic skills which students need in order to get good jobs when they leave school'.

Hand in hand with the discovery of new methods is the need to recognize the diversity of mental abilities as a basic fact of heredity–environment interaction, and to develop selection tests which spot the individual who is a visualizer as well as the one who is a verbalizer, and the imaginative, creative individual as well as one who has a quantitative analytical mind. To be specific we need to invent a battery of tests that do not correlate, or show the minimum correlation, with the results of conventional IQ tests.

When should these tests, once devised, be applied? There are two points here: the discovery of special abilities and aptitudes and the demonstration of general intelligence. The latter, Burt suggests, can be discovered soon after a child has entered school at the age of five, but special abilities probably do not emerge before the age of nine or ten. Before this the influence of these abilities is so slight that, unless a survey covers thousands of individuals, the evidence is swamped by errors of measurement and sampling.

What seems to be important is that an assessment of a child's ability should be made, not once and for all at the age of eleven, but right up the age scale. This arrangement would provide for refined selection with re-selection from five to twenty-one and could correct the errors of previous tests and bring to light other developing abilities. On this point we still know hardly anything about the interaction of genes with pre-natal environment and their interaction with later environments or even their interaction with each other.

The emergence of special gifts during puberty stresses the importance of continuing the study of each individual child right through his school life. Burt writes that such studies will reveal not only specific abilities but also specialized *disabilities* which, when rectified, could transform a backward child into a normal scholar. The recognition, and correction, where possible, of inherited defects like short-sightedness, astigmatism, deafness, and defects in colour vision can work wonders in learning. And other less simple inherited traits, such as differences between children in visual and auditory imagery, need to be discovered. Defective memory provides one of the commonest explanations why children of quite high intelligence at times seem educationally sub-normal; and on the other side of the coin, dull children with retentive memories may at times do well at the mechanical work of the classroom. Once these special disabilities have been detected and appropriate changes made in the teaching methods, the 'backward' child can often be transformed into a perfectly normal scholar.

It is sometimes urged that special schools should be set up for boys and girls with special talents—scientific, mathematical, musical, artistic, and so on. Whether we do this or not depends very much on how much these abilities are linked to each other or independently inherited. Burt shows how difficult this is to determine with eleven-year-old children, by citing the example of dividing a random sample of a hundred eleven-year-olds into three equal groups by tests to decide those who are good at grammar school work, those who are average, and those who are poor at it; next to reclassify them into those who are good at technical work, those who are average, and those who are poor at it. Out of the thirty-three who are good at technical work, about twenty will probably be good at grammar school work and only three definitely poor at it. In other words, Burt says, success at this age still depends far less on special aptitude than on "all-round mental efficiency which manifests itself to a varying extent in nearly every type of subject'.

The idea of such special schools is, of course, partly based on the needs of individuals, but there are also the demands of society for a flow of trained manpower. Clearly the uniqueness of the individual will lead to some incompatibility between his needs and those of society. Many of the needs of the individual are compatible, but we should think very carefully before forcing individuals into jobs or types of school incompatible with their personality and intelligence.

It is worth stressing here that great advantages would accrue to the community and to civilization generally if geneticists, psychologists, educationists, teachers, and medical experts could together systematically assess the potentialities of each individual throughout his school life and eventually fit the job to the man, and not, as is usual now, the man to the job.

Concern about the uniqueness of the individual and its survival in our society is obviously felt by those who are fitted to recognize it. This fact is emphasized in three quotations. The first is from the Newsom report, worth repeating here:

> There were well over two and three-quarter million boys and girls in maintained secondary schools in 1962, all of them individuals, all different. We must not lose sight of the differences, in trying to discover what they have in common.

The second, by the geneticist J. B. S. Haldane, also stresses the importance to society of individuality:

> That society enjoys the greatest amount of liberty in which the greatest number of human genotypes can develop their peculiar abilities.

The third is a statement by the contemporary geneticist, Thoday, on the aims of education:

> The aims of education may be divided into two groups, those concerning the needs of the individual, and those concerning the needs of society. Individuals vary in their capacity to respond to different modes of education in all these aims. Individuals have to be taught and vary in capacity to learn
>
> 1. to look after themselves
> 2. to get on with others who are different from themselves
> 3. to acquire those skills that will maximize their potential contribution to society and hence their success in life
> 4. to acquire appreciation of those of the good things of life that will make life rewarding for *them*
> 5. to develop their peculiarly individual creativities
> 6. to develop their limited critical faculties as far as possible so that their gullibility or exploitability may be minimized.
>
> Society needs
>
> 1. individuals who can look after themselves
> 2. individuals who can live with one another which of course requires that the individuals recognize that others may have different needs
> 3. individuals with vocational skills in frequencies proportional to the needs of existing social system
> 4. the transmission of tradition so that society may have continuity over time and may profit from the accumulated experience of earlier generations
> 5. individuals of critical and creative ability so that society may change, and tradition shall not become a dead hand on change. Such critical ability must be such that it is able to distinguish between traditions that are sound and should be preserved and traditions that are ill-founded and needing change. Change merely for the sake of change is not our need.

9: Chromosomes, Genes, and Gaols

SO far we have seen that intelligence, one important dimension of personality, is strongly inherited.
What of the more unpleasant side of human personality, whose expression can sometimes land a man in gaol for murder, theft, or rape, and can, as we all know, cause a man (or woman) to become one of the common menaces of the age—the aggressive, horn-blowing driver, who hates to be overtaken and helps to swell the accident and road death statistics each year? The question we are asking is: 'do criminality and aggression, like intelligence, have an inherited component?' Or, to ask a larger question about human personality: 'does heredity account for any of the differences and similarities between individuals in matters of human *conduct*, especially those that can land a man in gaol on the one hand or make him a pillar of society on the other?' If firm evidence can be provided for a large hereditary component in such traits as aggressiveness, instability, or irresponsibility (components of 'criminality') then free-will (actions independent of any cause) is in some doubt. To put it more dramatically, as Galton did: if heredity does influence conduct we are on the 'continuous merciless march of the hidden weaknesses in our nature through sickness to death'.

The large and important questions asked above are partly answered by three pieces of research: a study of criminal twins, chromosome abnormalities, and psychological studies of the personality of criminals and aggressive drivers.

CRIMINAL TWINS

In his book, *Crime as Destiny* (1931), Johannes Lange supplied the basic answer to these questions with a great deal of success. In brief, he demonstrated that criminal behaviour is very powerfully based on genetic factors and that environmental factors are relatively weak, though, as with intelligence, genotype-environment interaction is important for the expression of criminality. Lange's conclusions are based on an examination of thirty pairs of twins, one-egg and two-egg types, whose importance to the understanding of human individuality we have examined in Chapter 2, for their criminal records. In essence, Lange used the twins to show the degree of similarity or not in their *moral* decisions. Twelve pairs of twins were of the one-egg (identical) type, sixteen were of the two-egg (non-identical) type, while the character of the other two pairs was in doubt.

Taking first the sixteen pairs of two-egg twins, we know from Chapter 2 that genetically they are no more alike than ordinary brothers and sisters, or 'siblings' as they are called technically, and, except in one case, only one of the two was a criminal.

For the pair who were *both* criminals, their crime record differed in the timing of the first offence and in the amount and type of crime. Their personalities and body build differed too. One was a 'miserable physical specimen', fat, stupid, dull and a vagrant, the other 'tall, thin, cheerful and temperamental'. The latter served his first sentence at the age of twenty-four, the last at twenty-six, for theft, swindling, and false information. After this brief entry into crime he became a respectable person and an office clerk to a local authority. The other got into trouble with the police repeatedly, first at eighteen and the last time at thirty-eight, mainly for theft and swindling but also for vagrancy, begging, and lying.

The evidence from the *one-egg* twins is quite decisive. Not only do the records of eight of the twelve pairs studied show that both were criminals, but the records agree closely in the timing of the first offence, in the type of crime, and in their behaviour in court and in prison. Lange's words, as vivid as Dostoievsky's, speak for themselves.

> The Heufelders are old burglars, both of whom have been behind iron bars for nearly two decades and both of whom show paranoiac symptoms in prison. Both brothers Meister commit puerile offences against the laws of property, and both in prison suffer deeply owing to their terrified imaginations. The Lauterbachs are quite unusual swindlers, crooks almost of genius, who keep the upper hand even in court, and whose 'respectability' in prison is almost as great as their unblushing impudence. Both brothers Ostertag have just too little sense and will-power, at least in view of the ambitions induced in them by their happy, prosperous youth. The two Dieners, guttersnipes, but good fellows at heart, cannot stand alcohol; it makes them rabid and draws the knives from their pockets. The Maat brothers have not a scrap of affection for anyone in the whole wide world except their own unpleasant selves. Their abnormal sexuality leads them into intimate relationships, but even these only seem to be of value to them if they can exploit those with whom they are involved. Finally, the sisters Messer suffer from a degree of nymphomania which must be rare. In all these cases we see the results of the common law which binds these pairs of twins to one another.

In fact, to put it bluntly, they show not the slightest evidence of free-will as to their criminal behaviour. If we examined the record of any criminal who had an identical twin, on Lange's results, we should be able to predict the behaviour of the co-twin when placed in a certain environment.

A link is missing, however, before we can be certain of the omnipotence of heredity. Much more evidence is needed about the criminality of one-egg twins brought up apart from very early infancy. As we have seen in Chapter 3, a correlation of 0.87 has been found by Burt between the intelligence of one-egg twins separated almost from birth. Later work than Lange's, however, does not destroy his theory but moderates it. On average these later studies, which do not draw on such highly selected material as Lange did, suggest that criminality occurs in both one-egg twins about three times as frequently as it does in only one member, whereas almost the opposite holds true for two-egg twins. Environment, of course, must be important and this is borne out by what is known of juvenile delinquency, when both identical and fraternal twins tend to be alike. Probably the genotype determines the traits that dispose individuals towards crime—aggressiveness and brawn—and given the set of environmental circumstances that bring them out, these traits become apparent. In short, as Haldane wrote in his introduction to Lange's book, a man of a certain genotype put in a certain environment will be a criminal.

CRIME AND ENVIRONMENT

Lange's twins help us to sort out the relative importance of the environment and heredity in the dispatch of crime. He suggests six categories of evidence that bear on the environmental components in crime. These are: accidents to the brain at birth; accidents to the brain when adult; the influence of companions; discipline; disease and poverty. Of these the first two are the most important of the purely environmental agents affecting criminal behaviour, the other four depend partly on the genotype of the individual. Some of the evidence is set out below.

The two pairs of Lange's one-egg twins where only *one* twin took to crime are worth examining. One of the first pair (the Balls) brutally murdered a girl who was alleged to be going to have his baby. His twin brother, who closely resembled him, grew up as a law-abiding peasant. In the second pair, the Hiersekorn twins, one was feminine in looks and behaviour and had been in gaol for homosexuality. The other was normal sexually. Why the differences? In the Ball twins, the murderer had been born with a swelling on his head 'as big as a hen's egg' which later vanished. This might have been responsible for the tendency to abnormal states which he suffered from and in one of which he did murder. In the Hiersekorn twins both had received injuries at birth. The normally-sexed one had a damaged shoulder, the homosexual suffered from facial paralysis on the right side, probably, says Lange, evidence of brain damage at birth. So the differing characters of these two pairs of twins could be due to external accidents at birth which altered the normal expression of heredity.

Of Lange's ten pairs of one-egg twins, two pairs, although both criminals, took up different patterns of crime. With the Schweizer twins, taken away from each other at the age of eight and remarkably alike, this was probably because they were both completely without will-power and 'became creatures of the environment'. Strict discipline in their youth kept them on the rails. One got excellent school reports for diligence and conduct under the stiff discipline of his school, and the other, living with his foster father, although irresponsible, did not go to the bad. After running away, however, he went completely to the dogs, led a wild life, and took to drink and women, and eventually did violence to a rival. Likewise his brother, who got in with louts, stole and embezzled. But this one, who had done well under school discipline, likewise had good conduct during military service and in prison. Their marriages too were decisive to their later behaviour because they were so easily influenced. One eventually married a decent, strong-minded woman who took him in hand and kept him from trouble (and no one knew better than she, Lange writes, that 'he would always be capable of going to the bad'). The other had poorer luck with his women and was continually in trouble.

The Kramer twins too were convicted for different crimes; one, convicted at the age of twenty for repeated wounding, had been hit violently on the head with a jug when he was eighteen and this may have damaged his brain to alter his behaviour, the other was convicted for theft at sixteen. In this case, then, a violent knock on the head in one twin may have shifted his mental and physical development in a different direction from his brother's. In the case of the Schweizers the influence of discipline and of a strong-minded companion, here a wife, is of importance, though the effect lasts only while it is being exerted. As Lange says 'their conduct is not determined by the environment itself so much as by their innate tendencies, which deliver them both up to whatever may be the stronger influence of the moment, be they good or evil'.

Disease can be a very important factor in influencing behaviour, though it is one that can depend on heredity as well as accidental infection. As far as crime is concerned, Lange discovered a pair of (one-egg?) twins of whom one had goitre and might also have suffered from encephalitis but the other was normal. The former, from very early in childhood, was fatter and quieter than the other and remained so. Temperamentally the goitrous twin was more serious, settled, and dependable than his twin, though not such a good talker. He soon found himself a permanent job. His co-twin was a swanker, more intelligent though much wilder. Eventually he was imprisoned for breach of trust, embezzlement, and forgery. His brother lives a contented happy life, 'a picture of complete reliability and friendliness'. While it is not clear that the two brothers are in fact identical twins, Lange claims that close inspection of them revealed very definite resemblances. Goitre is a disease which understandably alters temperament and this may account for the marked difference in the behaviour pattern of twins. It is reasonably clear that the tendency to goitre is inherited (see Chapter 18).

Poverty is another environmental factor responsible for the development of criminality, and, of course, it is important, but probably only for some individuals. With poverty goes a multitude of other conditions which might act in different ways on different individuals: bad housing, large families, father absent or in prison, mother sleeping around, low intelligence. Poverty (and wealth) acts in strange ways and illustrates the principle described earlier in Chapter 8 of unique genotype interacting with unique environment to produce unique phenotype. Wealth may foster idleness, drunkenness, and swindling in one individual, and high-minded public service and philanthropy in another. Poverty may cause one to steal, another to better himself, and a third to continue poor but honest.

THE MARK OF CAIN

Even though studies of twins are important in the understanding of the inheritance of criminality, they are perhaps not so telling as the fact that criminals crop up in families free of criminality. This is powerful evidence that some are born with genetic outfits that pre-dispose to crime, just as we saw in Chapter 4 that ordinary parents can have extraordinarily talented children by genetic recombination. Lange's results show this. Taking the ten pairs of one-egg twins with a similar crime record, only four out of forty-seven close relations (parents and brothers and sisters) of the criminal twins were convicted, and the four were in one family. Of the fifteen two-egg twins, in every case one of the pair was sentenced and the other kept clear of the law. Clearly here is heredity at work dealing out vices and virtues—or rather the genetic traits that go to create them. How else can we explain criminals who crop up in a crime-free family?

We do not know much about genotype–environment interaction in relation to criminality, though we have the evidence of the weak-willed Schweizer twin. The effect of discipline in youth and a strong and steady wife caused him to go straight while the influence was exerted; his phenotype in respect of criminality was different when these effects were in operation, compared with the criminal phenotype of his wilder days of the knifing episode. The subtle interaction of genotype–social environment in respect of criminal behaviour is considered further in the story of the extra Y chromosome in the next chapter.

10: Extra Chromosomes and Criminality

THE weight of evidence that we are assembling in this book has shown so far that intelligence, educability, and certain qualities of character appear to be controlled by the interaction of numerous genes with each other and with a vast range of environmental factors.

Genes have never actually been seen down a microscope. Their presence is inferred from the results of a mass of experimental work with flies, microbes, plants, and animals, and observations on human family trees, twins, and so on. Some of these 'elements' of heredity or genes can now be put on a chromosome map in many organisms, including man. This map shows them to be in fixed linear orders, and it seems reasonable to assume that they are lodged on the chromosomes as stated in Chapter 1. When the gene outfits are suspected of playing a part in criminality, as in Lange's identical twins, or in the making of some modern genius, we cannot see the genes and we do not know their position on the chromosome map. The human chromosomes, however, can be seen down the microscope and can be easily counted. This fact has helped to link certain chromosome abnormalities with crime, thus adding to the evidence that our genotype powerfully influences personality and behaviour.

THE EXTRA Y CHROMOSOME

In 1965, Dr Pat Jacobs and others working with her in Scotland published a paper showing a link between the possession of a chromosome abnormality (forty-four chromosomes plus an X and two Ys) and a syndrome of which the main features among *hospital* patients are:

1. personalities which showed extreme instability and irresponsibility with severe impairment of the ability to consider the consequences of their actions;
2. in *general* low intelligence (IQ 70-90);
3. tall stature (183 centimetres compared with those with normal chromosomes, 170 centimetres) but no physical abnormality;
4. a remarkable absence of crime records and extra Y chromosomes within the families of the XYY when, compared with other patients at the hospital in question, surely an indication of the importance of inborn factors;
5. a history of convictions since boyhood, mostly for offences against property but some against persons.

Such XYY men have been found in prisons and borstals in Scotland, Australia, America, and England. In 1970 Wandsworth Prison, for example, contained 9 XYY men (all over 180 centimetres tall), out of 355. Pentridge Prison in Australia contained 5 such men (over 183 centimetres tall) out of 40 prisoners. In short, it looks as if the XYY chromosome outfit gives its possessor a greater risk of ending up in gaol; however, it has been suggested that such men have only three to four times as much risk as a normal man of ending up behind bars in Great Britain.

It seems incredible that in England and Wales (and probably most other countries), one in 700 male babies is thought to be born with the XYY 'karyotype' as it is called. This means that there must be in England and Wales about 40 000 XYY males roaming around. Why is it that some end up in prisons and special security hospitals while the large majority remain free, since there is evidence that to have an extra Y chromosome can predispose a man to behaviour disorders?

We are surely back here to the genotype–social environment interaction and thus must look closely at the background from which these imprisoned XYY males come.

Perhaps an extra Y chromosome does not in itself load the dice towards delinquency, but only in combination with low IQ and poor family background: in other words, an extra Y chromosome *by itself* may not be such a major catastrophe for an individual. In this respect it is interesting to see where these hospitalized men come from. Most come from families of skilled, semi-skilled, and unskilled workers; most from backgrounds of large families where there are illegitimate children. And, significantly, a third come from homes where the father or mother died or vanished before they were six years old. These conditions provide the soil where delinquency thrives in any case, but if, in addition, a boy is driven by an imbalanced set of chromosomes and genes that produce low intelligence, fate is loaded against him. What we must be wary of is the general statement that 'XYYs tend to criminal behaviour'. The statement needs to be qualified by specification of the particular social environment from which an XYY man or boy comes. As yet our knowledge of the backgrounds of 'normal' XYYs (unknown to the courts) and those who end up in gaol is not sufficiently precise.

SOCIAL IMPLICATIONS OF EXTRA Y

Of course, the statement we have discussed above that XYYs tend to criminal behaviour might alarm most people for its implications which would be that chromosome and biochemical tests may be desirable for all new-born babies, not just to pinpoint this particular abnormality but for many other chromosome abnormalities and genetic defects. No suggestion is being made here that XYY people should be restricted in any way before they have committed a criminal act. If every dog can have its bite, a genetic deviant must have his. But crime might be reduced in others if we could know more about the natural history of crime: a greater knowledge of human chromosomes and genes and inherited personality patterns which can be measured and correlated with future conduct and detailed case histories of the ecological background from which a delinquent or criminal has come. The geneticist K. McWhirter has suggested, as a first step, a way of treating the problem of criminality on a scientific basis. This is that all dangerous convicts should be graded on a G-E scale in which an estimate is made of the ratio of genetically (G) and environmentally (E) determined causes of their condition.

Those with a greater E component should in theory be easier to reverse. Those with chromosome defects such as an extra Y chromosome will not, because they are at the extreme G end of the scale. Special conditions might be created for an XYY man to ensure that he is *less* at risk (and if he is discovered early on in childhood the special environment could start early). The trouble is that rehabilitative measures designed for the average prisoner in whom there may be a larger E component may well be useless in the case of an XYY man.

Recognition of a high G component, be it imbalance of chromosomes or 'bad' inheritance generally, as in Lange's twins, is vital for the protection of the public, for segregation from the normal population is essential. McWhirter says, 'thus we have to plan for the setting up of strictly isolated communities where lives can be lived as usefully as genes (and the current stage of medical knowledge) permit.' In theory, perhaps so. But concentration camps and ghettos might be a much greater danger than the relief they might give. And there is an additional point that many genetic defects (phenylketonuria) are very easy to reverse; many environmental ones are irreversible.

11: Is Conscience Inherited?

ALTHOUGH we can marshal plenty of evidence from genetics of determining factors of intelligence and criminality, we have as yet left a blank page about the mechanisms—nervous structures or chemical hormones—which may underlie the kind of personality which lands up in gaol. It is this gap between the work of Lange and the new work on extra Y chromosomes and criminal behaviour that we must now try to bridge. Many of the facts and theories reported here stem from the work of psychologists like H. J. Eysenck and M. Parry.

THE CHARACTERS OF GAOL-BIRDS

The first point worth examining is whether men in gaol have a particular kind of personality or, more commonplace, whether drivers who are accident-prone through dangerous driving are special types. If Figure 7 is examined, the four temperaments derived from the ancient teachings of Hippocrates and Galen—choleric, melancholic, phlegmatic, and sanguine—can be seen in the centre circle. Outside this the results of modern research on personality traits are set out, and

FIGURE 7 Diagrammatic model of personality organization. In the centre are given the four ancient temperaments; on the outer ring are shown the results of large numbers of modern experimental and statistical investigations into the relationships obtained between traits. (From Eysenck, 'The Biological Basis of Criminal Behaviour', *Advancement of Science*, **21**, 1965)

two axes divide the circles into quarters. One axis, running from north to south, passes from the stable to the unstable personality and this we can call the 'emotionality' or 'neuroticism' axis. The other, the east to west axis, runs from the extroverted to the introverted personality type. What do these terms mean in everyday language? 'He's neurotic' often describes a man who is moody, anxious, and touchy, and flies off the handle easily, quite the opposite from the calm, even-tempered, if duller, person, the stable type. The traits linked with an extrovert nature can be seen in the outer circle of Figure 7 —outgoing, sociable, talkative, and optimistic—and those associated with an introverted character—thoughtful, unsociable, quiet, and careful. Most people, of course, do not belong to these extreme types but are mixtures, though they tend to go in one direction, obeying the ancient personality types of Hippocrates.

Criminals in gaols mostly fall into the choleric quadrant (neurotics in mental hospitals in the melancholic quadrant); that is, high on extroversion and instability (neuroticism). In other words people in gaol are often aggressive, restless, and excitable. It is the same with the driver who tries to edge us off the road, or shouts and swears at us and blares his horn; his personality traits are very much like those of the man in gaol—he is aggressive and anxious.

The prisoners in gaol are very like 'psychopaths' in their personalities. These are mental hospital patients who are characterized by a 'lack of conscience', who tend to be impulsive, have little concern over the rights and privileges of others, and accept little responsibility for their failures. Eysenck believes, on good evidence, that conscience, the mysterious quality that makes cowards of us all, is but a conditioned reflex, subject to the laws of heredity, and the study of some simple behaviour patterns may throw light on this matter of conscience and also on the possibility that heredity plays a part in its quality.

THE CONDITIONED REFLEX

The technical expression needs some explanation but refers to a commonplace everyday experience which involves transferring responses from one stimulus to another new one. A baby transfers his welcome from his mother's smiling face to the food and warmth she is associated with, to a song she may sing, or to a particular scent she may wear, and so on. Pavlov put this ordinary observation into scientific shape with his dog experiments. If tasty meat extract is put in a dog's mouth, it makes saliva by a 'reflex arc'. A reflex arc is a simple hereditary (unlearned), only slightly adaptable response (the saliva) to a stimulus (the meat) carried out through a chain of nerve cells. The thinking brain cannot alter these reflexes very much. To return to the dog story again; if a bell is rung at the same time as the dog has his meat, something happens in the dog's brain so that his mouth waters for the bell without the meat. In other words he has transferred his response from the unconditioned stimulus (the meat) to the conditioned stimulus (the ringing bell). All 'conditioned' means

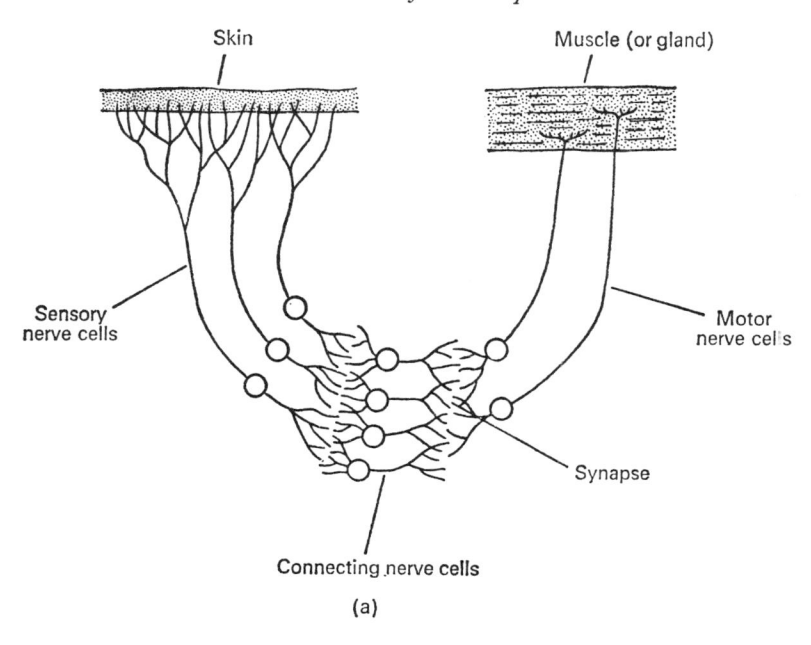

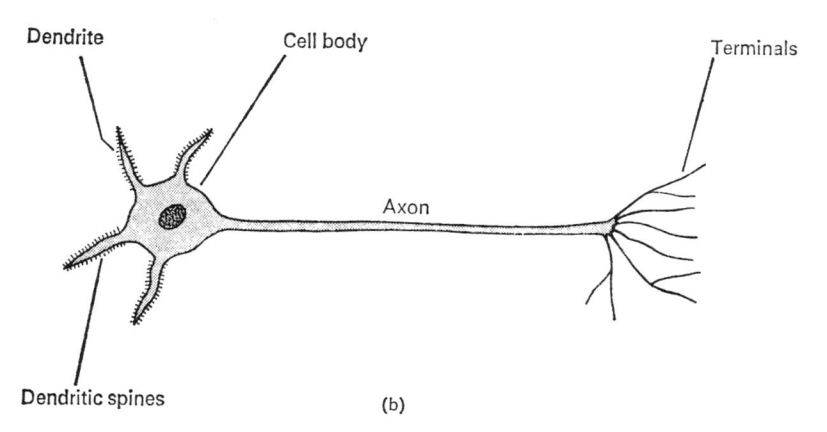

FIGURE 8 (a) The mechanism of reflex. The diagram shows a sense organ (the skin), sensory nerve cells which influence connecting nerve cells in the central nervous system, and motor nerve cells which bring about muscle contraction, or glandular secretion. All ordinary responses involve hundreds or thousands of nerve cells at each stage, usually linked in far more complex ways than those shown.
(b) A simplified diagram of a neurone.

is that the dog has been brain-washed into salivating by transferring the response, salivation, to the new stimulus, the bell. But what has happened in the dog's nervous system? The dog has got to start with something, in this case the unconditioned stimulus of the meat. There appear to be inborn nerve patterns in the brain which respond without experience to basic stimuli, the dog to meat, a baby to the shape of an oval with marks for nose, eyes, and mouth. (See page 124.)

In the transfer of response from meat to bell, Pavlov argued that two webs of nerves are involved, one leading from taste buds on the tongue into the spinal cord and brain (the sensory nerves) and out again to the salivary glands by the motor nerves (a reflex arc, see Figure 8(a)), and one from the ear to alerting centres in the brain. These two lots of nerve webs are not like separate telephone wires but must make contact at numerous points in the brain through nerve junctions called synapses. The junction is not continuous between two nerves but has a sub-microscopic gap in it, and this physical break in continuity between two nerve cells is bridged by a flood of chemical substances (transmitter substances) liberated by the nerve endings at the synapse in response to an electrical impulse travelling along the nerve fibre. Where the two nerve webs make contact, the synapses at first must be very resistant to nerve impulses. They have, we say, high 'thresholds'. But if both webs, the bell nerve web and the taste bud web, act at once, their impulses 'sum' at the common synapses, fire the bridging chemical substance and, at each firing (discharge of an electric pulse), lower their threshold so that the impulse gets across easily. Just how these synaptic thresholds are lowered is a current problem of physiology. After several lowerings the bell alone can activate the salivation path, but if no meat is given the thresholds drift back to a height probably set by heredity. This is the important point. It has often been shown in animals and humans that the speed with which conditioned reflexes are formed, the strength of the reflexes so produced, and their resistance to 'extinction', differ profoundly from one person to another or from one dog to another. Some people condition quickly, strongly, and lastingly (neurotics) and some condition only with difficulty (criminals). These differences between people reflect a predisposition on the part of the nervous system to mediate the establishment of conditioned responses well or poorly, and this may be the genetic basis of criminality.

CONDITIONING AND LEARNING

Before explaining this last statement, the difference between learning and conditioning should be made clear. Conditioning, as we have seen, proceeds through association, and usually what is acquired by conditioning is not so much a specific response as an emotion like fear. Learning, on the other hand, proceeds through trial and error, reward and punishment, and benefits from teaching. It is the central nervous system, the spinal cord and brain which are involved in learning. The brain sorts out the sensory input from eyes, ears, skin, and other sense organs and appropriate (voluntary) action is initiated through the motor system (motor = physical movements)—nerves which activate the skeleton muscles. In between most of the sensory and motor neurones is a third rank of nerves, the connecting neurones. The evolution of this intermediate network of nerves has been very important in man, in whom they form 99.95 per cent of his nervous system. Sensory and motor neurones form a tiny fraction of the whole. A rough calculation, for example, shows that there may be 2000 of these connecting neurones to one motor neurone in man. Their importance in the brain might be to select, combine, and direct messages for more elaborate processes—like turning a single piece of information on the retina into a face—or to store memories at their synapses. In short, they may be used for thought.

Besides the central nervous system, we have the autonomic nervous system, motor nerves which regulate bodily activities not entirely under 'the will': heart beat, digestive system, blood vessels everywhere, the tiny muscles that expand and contract the pupil, and many other systems. In brief, it takes over a load of routine from the decision-making centres of the brain and gets on with its activities without our having to bother about them. But we must be careful not to oversimplify, for, as stated in Chapter 18, we can choose to think about exciting experiences or certain other events and alter our emotional state. So here is the autonomic system acting under the direction of 'the will'.

The autonomic system exerts an on-off control; it can, for example, quicken or damp down the heart beat, and it does this by its two opposite components: the sympathetic system which gets the body ready for fight, flight, or fright, and the parasympathetic which returns the body to peace. In terms of chemistry this means a stable body temperature, water content, blood sugar, and many other factors. In brief, the autonomic system is closely related to the expression of the emotions. The two systems described are balanced just as is a cyclist who keeps upright even over rough ground by pressing one handlebar and then the other. Although it is true that the autonomic system is primarily involved in conditioning, we must remember that it is connected with, and is loosely under the control of, the central nervous system.

How does all we have said link up with the conscience or lack of it in criminals? Supposing a little boy punches his sister or, instead of asking to go to the lavatory, dirties his pants rather than spoil his game, his mother will smack him or punish him in some other time-honoured way. In this case an anti-social activity is immediately followed by a strong pain-producing stimulus. The conditioned stimulus is the 'naughtiness' in which the boy has been indulging, the unconditioned stimulus is the slap (or other punishment), and the response is the pain and fear produced in the child. In this situation conditioning should take place, so that from then on the particular type of activity would be followed by a conditioned (transferred) fear response. After a few punch-smack, dirty pants-smack situations, the fear response should be sufficiently strong to keep the child behaving according to the accepted social patterns. Of course misbehaviour is not always physically punished but is labelled 'bad', 'naughty', or anything else the parents choose, and the child groups all these 'bad' activities together as being potentially punishment-producing and productive of conditioned anxiety and fear responses. As a burnt child shuns fire, so a punished child shuns wickedness.

What about temptation to do one of these forbidden things? The child will of course tend to go and do it, but as he approaches the object of his temptation—say taking money from his mother's purse—there should also be a strong upsurge of the conditioned emotional reaction, fear or anxiety, which is associated with such activity as stealing money. If it does not, he will carry on. As Eysenck says: 'Whether he does or does not behave in a socially approved manner depends essentially on the strength of the temptation and on the strength of the conditioned avoidance reaction, which has been built into him through a process of training or conditioning.' But whether he conditions fast and lastingly depends on the inherited quality of the autonomic nervous system.

The growth of conscience in a child depends largely, of course, on the environment he is brought up in, and the moral and social standards of parents and teachers. But a child whose nervous system has the inherited tendency to develop conditioned responses slowly and poorly has the tendencies of the psychopath, the criminal, and the extrovert. On the other hand, the neurotic, the law-abiding citizen, and the introvert generally, may be characterized by strong and lasting conditionability. Some experiments on puppies by the American psychologist Solomon will help us to understand this.

THE CASE OF THE CRIMINAL PUPPIES

The experiment was conducted with six-month old puppies that had been starved for two days. Two dishes, one filled with boiled horse meat and the other with commercial dog food, were placed in the room with a chair for the experimenter to sit in. The puppies mostly went straight for the horse meat, but as soon as one touched it he was swatted with a rolled-up newspaper, and if one gentle tap was not enough the puppy was swatted again and again until he gave up and turned to the commerical food which he could eat in peace. This training was carried on for several days and then the two dishes were offered in the absence of the swatting man. The choice had to be made between the tinned food and the real thing. The puppies gobbled up the commercial food and then began to react to the dish of horse meat. In Solomon's words: 'Some puppies would circle the dish over and over again. Some puppies walked round the room with their eyes towards the wall, not looking at the dish. Other puppies got down on their bellies and slowly crawled forward, barking and whining. There was a large range of variability in the emotional behaviour of the puppies in the presence of the tabooed horse meat.' Resistance to temptation was measured as the number of seconds or minutes which passed before a puppy ate the tabooed food.

The agony of conscience continued. The puppies were allowed half an hour a day in the experimental room. If they did not eat the horse meat by that time, they were put in their home cages but not fed and a day later were again introduced into the experimental room. This continued until the puppy finally broke the law and gobbled up the horse meat or alternatively, if the puppy had been so conditioned that he would rather die than eat the meat, he had to be fed to keep him alive. The longest period was sixteen days without eating, the shortest six minutes. Two points are interesting.

First, the great range of variability in the puppies: the law-abiding ones who would rather starve than violate the taboo and whose very strong 'conscience' was created with but a few swattings, and the criminal puppies who broke the law without putting up much of a resistance. The reason for this is the very marked difference in conditionability, which depends upon inherited differences in the speed with which the conditioned reflexes are formed, their strength and their resistance to extinction, this latter depending on how quickly the nerve synapses return to their original thresholds. The criminal puppies were born with a nervous system genetically predisposed to develop conditioned responses slowly, the law-abiding ones with nervous systems which could be conditioned strongly and with ease.

Second, a point which might show the importance of close family relationships in the growth of conscience: when puppies were hand fed throughout early life they developed a conscience much more strongly than did other puppies who had been machine fed, and this implies that young organisms learn better if they get some love.

INTROVERTS AND EXTROVERTS

In Figure 7 and throughout this chapter the polarity between introverts and extroverts and the link between their behaviour and the conditioned response have been underlined. The basis of the difference between extroverts and introverts, judging from studies on one-egg and two-egg twins, appears to have a genetic component. For introversion one-egg twins have a correlation coefficient of 0.50, two-egg twins of between 0.40 and 0.20. It seems reasonable to suggest that the effect of the genes is on the strength or weakness of the conditioned response, and this in turn influences the growth of conscience. The latter, of course, depends on the actual application of conditioned and unconditioned stimuli by parents, teachers, brothers, sisters, and others. At the start of the chapter we noted that criminals showed not only a high degree of extroversion but also a high degree of emotionality or instability. Where does this fit in?

Most important is the fact that this quality of emotionality is closely related, again to the autonomic system. Some people's blood gets up very quickly—their emotions are produced very quickly, strongly, and lastingly—and they are prone to over-emotional, unstable conduct. By contrast other people are very even-tempered; their emotions are difficult to arouse, are not strong, and are short-lived.

In between come the large majority of us. Like extroversion–introversion, emotionality has been shown to have a large hereditary component, by studies on animal, and also by studies on one- and two-egg twins; the former having a correlation coefficient of about 0.60, the latter of between 0.30 and 0.20. How does extroversion–introversion mix with emotionality to give criminal behaviour, given that there is a strong hereditary component at the basis of each quality? In brief, emotion, and particularly strong emotion, acts as a drive. This added to a high degree of extroversion can produce the formula for criminality. Imagine a Bunsen flame burning high, yellow, and billowy, with the air hole closed (extroversion). Open the air hole and the flame roars into a blue, hot jet.

Strong emotion (the air) plus extroversion can make a man into a criminal just as the flame can change in quality. Likewise strong emotion plus introversion can cause a man to become a neurotic. Criminals and neurotics are, then, doubly predisposed by genetic factors; they condition too poorly or too well, and their resulting 'habit' systems are boosted by their inherited strong emotions to give the violent behaviour leading them into prison on the one hand or mental home on the other.

THE SWEARING, HORN-BLOWING DRIVER

Quite apart from the extremes of criminality and neuroticism, there are the twin demons of aggression and anxiety in a great many of us. They often show themselves when we get behind the steering-wheel of a car. The psychologist M. Parry found that male drivers with high aggression and high anxiety have by far the highest accident score. Among women, too, the high aggression-anxiety group is the most accident-prone.

Who are the aggressive types? The people with the highest aggression scores (based on a questionnaire containing such a question as: if another driver makes a rude sign at me I do/do not do something about it) tend to be men aged 17–35, particularly those below 25. Aggression scores fall off steadily with age. Anxiety is more complicated (I feel a little more apprehensive/I never feel apprehensive when I notice a police car about or following me, was the kind of question Parry asked to determine the anxiety score). Among the men it falls off with age but among women it rises.

Parry found in his survey of 382 drivers (279 men and 103 women) that 27 had driven at another car, 45 had tried to edge another car off the road, and 32 had been in fights. Here is one story told to Parry by a 28-year-old accountant, in the high anxiety-aggression group.

'I remember once, about six months ago, being followed by this bloke with his car headlamps full on. I thought at first he wanted to overtake me so I pulled over to one side to let him pass—he didn't. In fact every time I allowed him to overtake he slowed down as I did. Eventually I became so annoyed I pulled up, thinking he would also stop. He carried on and, as his car passed mine, I noticed three blokes, all about twenty or so, were in it. They hooted at me, gave me the V sign and drove off. I was pretty annoyed so I chased after them and gave them the same treatment. It developed into a running battle and at one time I drove up alongside and gradually crowded their car into the dirt. I was so angry I would have gone on crowding them right off if I hadn't noticed a police car behind.'

This is a relatively minor violation of the moral code of our society but the man concerned is part of the genetic spectrum which shades into criminality. Indeed there is a good deal of evidence that known criminals commit a greater number of traffic offences and have more serious accidents than do non-criminal citizens. G. Willett found in a sample of 653 road offenders in Great Britain that four out of every five who caused death by dangerous driving had criminal records; of 69 who drove when disqualified, 78 per cent had criminal records; of 104 driving when drunk 20 per cent, and of 285 driving dangerously 15 per cent, had criminal records. This earlier British work has been taken further by F. A. Whitlock, in Europe. The maps in Figure 9 show that a European country with a high rate of murder, suicide, and other violent deaths will have a corresponding high rate of road deaths. One map shows road deaths, the other how people die violently by other means. West Germany, Switzerland, and Austria are in the top category in both maps; Spain and Ireland in the bottom category. Whitlock suggests that road accidents can be used as a sensitive barometer to indicate the amount of pent-up violence and aggression within a society. It is odd at first sight that Switzerland, a supposedly orderly and peace-loving country, is top of the League table, and Spain, where death and violence seem common-place, is at the bottom. Whitlock speculates that lack of external threats in Switzerland has something to do with internal violence; Spain might have had its aggression drained away during the Civil War of the 1930s. But we

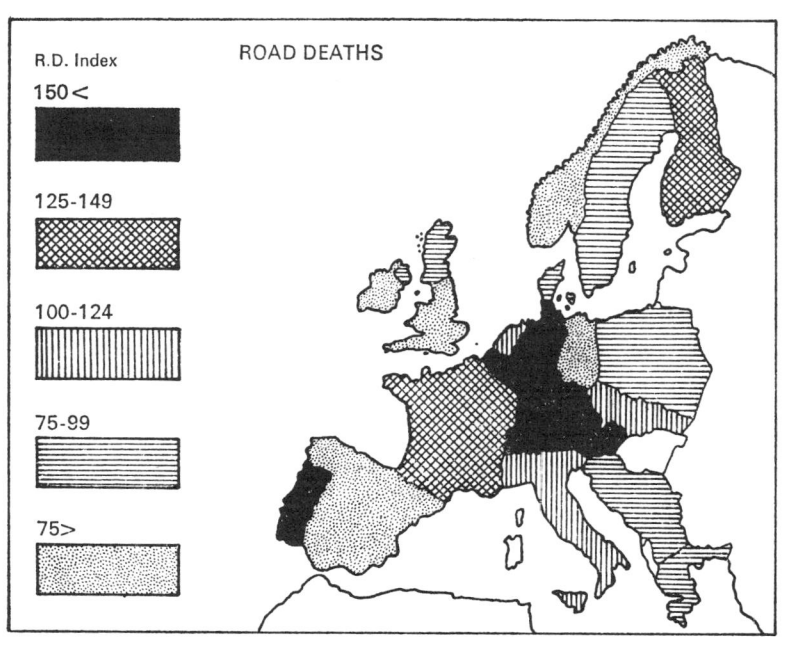

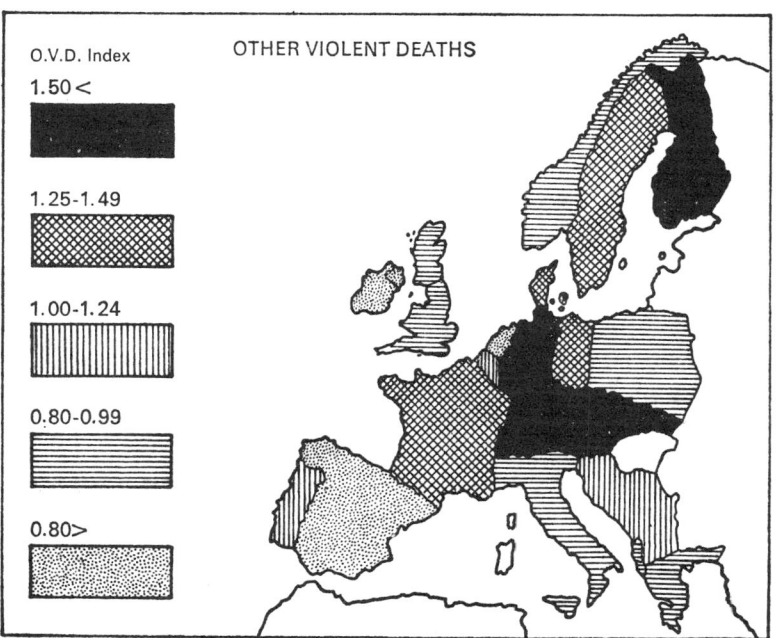

FIGURE 9 Road deaths (left), averaged out, compare closely to patterns of murder, suicide, etc. (From *Sunday Times*, 20 June 1971.)

need to take into account geography and winter climate in the interpretation of these data: Switzerland and Austria have bad winters!

ENGRAMS

It would not do to leave this chapter without the obvious reminder that the cortex of the human brain is a wonderful instrument of intelligence and simply does not operate by 'mechanical' reflex arcs, although the rudiments of human behaviour can be explained on a basis of conditioned reflex, as we have seen. Indeed the criminal and the vicious driver appear to be doubly disposed by heredity to commit crime when put in the right company and conditions, but each makes a personal decision about his crime as a result of past experience and present information. What is being said is that most human conditioned behaviour is quite opposite to reflexes; it is a personal, learned, intelligent response by the brain cortex to a given situation. Reflex arcs are mechanical. We shall look at brain function in more detail in Part 2 but one example of this individual, conditioned, learning might clarify the argument.

When a baby is born it has no experience, and like the dog reacting to meat it has to start with something. This 'something', discussed more fully in Chapter 15, might be an inborn pattern of nerve cells in the brain which are keyed to respond to basic stimuli, a face, especially a face with a mouth curved into a smile. The baby's pleasure is reinforced by food, warmth, and so on. A face-pleasure-food link is thus formed. But now other patterns are woven on to this one through nerve connections, which have at first high thresholds but which with continual use become a 'nerve web', so the face has become 'mother' with her smile, warmth, scent, and food. These webs are called engrams (a kind of writing left behind in the brain by conscious experience) built up, perhaps, through nerve connections in the cortex by learning. Our brains, like our finger-prints, are each individually tailored from the patterns laid down by the genes and by the connections formed by the dendrites. So the kind of engram each of us weaves through conditioned reflexes is very personal because of unique heredity and unique experience and is in complete contrast to reflex arcs. Perhaps the brain is one huge engram, because each experience is interwoven through the nerve webs with another. What we do not know is why similar engrams produce different qualities in different people, nor do we understand creative drive, or curiosity. But, as H. C. Elliott says in a masterly work on the brain and intelligence, 'these things are as surely biological as are reflex, instinct and transferred (conditioned) response'.

12: Angst

ANXIETY is a common state of mankind, and no wonder with all the stresses of city life and living in a competitive society. Anxiety can, as we have seen in the last chapter, add a disastrous element to human personality in combination with aggression. We all recognize the very anxious person in everyday life: often conscientious to a degree and suffering from a profound sense of time urgency. He may have health panics, is a worrier, and is often shy and sensitive. When out in the car he may imagine himself in the centre of a nasty accident and he hates night driving. He often explains himself rather forcefully and rapidly with accompanying sudden gestures. The anxious person often has a 'lump' in the throat, a knot in the stomach, and has difficulty in going to sleep because of the thoughts going round in his head. We all know too that anxiety is catching. When in the presence of this very anxious person, it is difficult to remain unaffected for long. Most of us are not as unfortunate as the person described, but all of us have some degree of anxiety. Indeed a moderate degree of it in response to stress is healthy; it concentrates the mind, gives us energy; in short it has survival value. Naval officers during the last war living a life of great danger were keyed up and tense—

anxious, but able to react energetically and effectively in an emergency. At sea the men felt well because there was something for their tension to grip. At home they flopped and had sleepless nights; the anxiety had continued and was useless.

What interests us in this book is whether this important element of human personality is something one is born with in differing degrees, making some people more anxious than others, and also whether anxiety can be explained in terms of chemistry. The answer to the question on inheritance is 'yes', genes do play a part in anxiety, but, like all the other aspects of inheritance, only a part. The development of 'angst' is probably governed by an interaction between the stresses of the environment and inherited factors. As for the chemistry of anxiety, although the work is in an early stage, the emotion can be plausibly explained by chemical changes in the body.

ANIMAL ANXIETY

Animal studies have helped a great deal to show whether anxiety is inherited, because it is possible to control both their breeding and environment in a way that is impossible in humans. As every dog lover knows, there are genetic differences in temperament between breeds of dogs. Some recent work with the pointer suggests that *within* breeds there is considerable genetic variation too. The best animal experiments have been done on the rat by Broadhurst. He obtained strains that became increasingly different in emotional response by repeated pairing of the most and the least emotionally reactive rats with their litter mates. The emotional response was measured by the number of pellets defaecated when the rats were exposed for two minutes in an unfamiliar, bright, and noisy, open arena. After fifteen generations of breeding further differentiation of the strains was still being achieved, a fact that suggests that many genes contributed to the genetical background of this behavioural trait.

STUDIES OF TWINS

So far as predisposition to anxiety is concerned, mankind seems to be as variable as the rat and the dog. Darwin noticed this in relation to emotion—blushing, for instance—and he postulated that the tendency to blush easily was inherited 'as a direct result of the constitution of the nervous system'. Since then many observations have been made which suggest there is genetic variation in physiological reactions related to anxiety. Changes in heart beat frequency and breathing rate, for example, following a sudden noise.

Conclusions about the inheritance of anxiety have come from the comparisons of likeness and differences in one-egg and two-egg twins which once again, as Galton put it, provide unique opportunities of 'weighing in just scales the effects of nature and nurture'. Most of these studies have been made not to measure anxiety specifically, but introversion and other characteristics of personality closely related to anxiety. Those that have been done show that 41 per cent of one-egg twins suffer from anxiety neurosis, as against 4 per cent of the two-egg twins. When anxiety is considered more broadly, it is found that when one of the twins suffers from marked anxiety, it is also recorded in 65 per cent of the identical (one-egg) twins but in only 13 per cent of the fraternal (two-egg) twins. As some of the anxious one-egg twins had been brought up apart it seems that the development of anxiety reactions to stress is the expression of a basic genetic endowment. The results just quoted come from some scrupulously fair tests by Eliot Slater and J. Shields at the Maudsley Hospital, where the cases of anxiety were blind diagnoses. Eliot Slater was presented with summaries of each case which did not say whether the other twin had been mentally ill or whether the pair were identical or fraternal. So bias was prevented.

A study of family trees, too, points to the conclusion that anxiety is inherited. Fifteen per cent of the parents and brothers and sisters of anxiety neurotics are similarly affected in the Maudsley study. One carried out at Harvard showed that about 5 per cent of the American population were affected, and where one parent had anxiety neurosis, half the children suffered from it too. When neither parent suffered but one child did, about a third of the other children in the family had anxiety neurosis.

None of what has been said denies the influence of the environment on the development of anxiety neurosis. In the Maudsley study quoted many of the identical twins were unaffected, and even when both members of an identical pair suffered from anxiety, their symptoms and severity could differ extensively. Again we have clear evidence of genotype-environment interaction to produce the anxiety state, just as interaction could make for criminality or ability.

ANXIETY AND THE THREE-CARBON ACID

None of this tells us much about what mechanisms, biochemical and nervous, might be responsible for anxiety, although we know that it has a genetic basis. One important clue to such a mechanism was picked up when it was noticed that symptoms of anxiety neurosis resembled some of those produced by physical exertion—pounding heart, breathlessness, headache, chest pain, tiredness, and many others. This led on to investigations on certain physical functions in anxiety neurotics, and it was found that, compared with normal people, anxiety neurotics reacted sooner to increasing levels of noise, light, or heat. Their handgrip could not be maintained as long and their breathing rate went up more in response to discomfort. In response to light exercise their pulse and breathing rate increased more than in the normal person, and they used the oxygen they breathed less efficiently and developed a higher level of lactic acid in the blood.

This last fact is one of great interest. Lactic acid is the normal end-product of the process by which body cells break down glucose (or glycogen, in which form glucose is stored) to yield the energy for work. When the muscles perform work, large amounts of glycogen are broken down to lactic acid in a number of steps, each controlled by a different enzyme. At each step in the breakdown a small amount of energy is released to form energy-rich phosphates which are used for work. In yielding up their energy they change to energy-poor phosphates to be enriched again. The lactic acid formed diffuses into the blood, is carried to the liver, and changed back into glycogen and stored until needed. All this goes on in the absence of oxygen. When oxygen is available the lactic acid is split down by enzymes into carbon dioxide and water, and part of the energy is used to reconvert lactic acid back into glycogen.

What is important for us here is that lactic acid seems to be a key factor in anxiety. Many investigations over the past thirty years have shown the rise with exercise in blood lactic acid to be excessive in anxiety neurotics. Now the interesting point is that anxiety attacks can be brought on in anxiety neurotics and some 'normal' people by injections of solutions containing lactic acid, but mitigated or abolished by injections of lactic acid with added calcium. The reason for the calcium-lactic acid experiment is that some of the anxiety symptoms are similar to those seen in hypocalcaemia, a condition where the level of calcium in the blood is low. Lactic acid has a weak power of 'binding' calcium into an inert form, making it useless in the body, and calcium is very important in the transmission of nerve impulses. In the experiment enough calcium was added to saturate the binding capacity of the lactic acid, so an injection of lactic acid plus calcium would presumably leave the calcium level in the subject's blood and other body fluids unaltered, because the lactic acid was already 'saturated' with calcium and would not accept more from the blood and body fluids. Not surprisingly, hardly any anxiety symptoms occurred when the patients were injected with glucose solutions. So here is an anxiety attack triggered by the specific addition of a stimulus, a chemical, lactic acid. The anxiety neurotics likened the effect to their worst attacks—'heart pounding, mouth dry, vision blurred, dizzy, headache'.

How is anxiety brought on naturally? A theory, but only a theory, worked out by N. F. Pitts is that anxiety symptoms might be induced by the binding of calcium by lactic acid. If this binding happens in the fluids at the surface of nerve endings, an excess of lactic acid could interfere with the rôle of calcium in transmitting nerve impulses. But what is the source of the excess lactic acid? Briefly, it may be because of over-secretion of a chemical arouser, or hormone, adrenalin, which is well known to stimulate anxiety symptoms as well as to step up lactic acid production. Chronic over-activity of the autonomic nervous system, as we have noted in Chapter 18, sets the body on a twenty-four-hour alert and one of its consequences is over-production of adrenalin from the adrenal glands. The anxiety neurotic, then, could be someone with a genetic predisposition to: over-activity of the autonomic system leading to chronic excess of adrenalin and in turn lactic acid; faulty metabolism of the body so that too much lactic acid is made; a defect of metabolism so that too little calcium is made for nerve conduction to function properly. Anxiety may, of course, be due to a combination of these conditions.

All this is only a theory, but one which is compatible with a new treatment for anxiety neurosis. A drug has been found which can block the formation of lactic acid from glycogen and also prevent adrenalin having its nerve-stimulating effects. This drug can reduce or eliminate anxiety symptoms.

As conscience may well be a pattern of conditioned reflex, so anxiety, which in excess is a burden or in moderation a spur to clear thinking and energy, may in the brain be the result of an excess of a simple acid whose key structure is three carbon atoms ($C_3H_6O_3$).

Part 2:
A View of the Brain

Irrespective of race and sex, the basic components of our brain are modelled to a pattern laid down by heredity. Each of us has a cortex, the human thinking cap: a cerebellum, the automatic pilot of the body; a complex 'old' brain—old in the evolutionary sense—deep in the middle of the skull, whose mastermind is the hypothalamus, the physical link between thoughts and expressed emotions; and a brain stem where life and death functions like breathing are controlled. In man, it is incorrect to think of the emotions being 'superseded' by thought. Rather they enhance his appreciation of his physical environment; they enrich and alter his brain rather than override it.

That this basic brain pattern is the responsibility of a balanced set of chromosomes is shown by the effect of chromosome imbalance. A small defect wrecks the brain's machinery and its intelligence. Yet despite the fixity of its gross structure, in its fine structure, the neurone, the brain is somewhat plastic. The number and arrangement of synaptic connections varies from brain to brain, as does its elusive memory store. What enters the focus of a man's attention is recorded in a memory store. What he ignores leaves little or no lasting record. What we attend to or ignore varies from person to person.

But what alterations are necessary in the brain to make permanent recording of experience and memory recall possible are problems of current research.

Much of the brain is ready for instant use at birth. The visual, auditory, and motor cortex are ready but other parts are uncommitted. The speech and perception areas are like blank slates ready for the inscriptions of experience, and what experience we get is a matter of luck, in some cases cruel luck, and depends on the environment provided by parents and kindred, an environment depending very much on heredity.

The old brain, whose hub is the hypothalamus, is a complex system yielding slowly to the tools of the scientist. Somehow it is involved in emotional behaviour, but its parts are so interlocked and small that to ascribe to any one a specific function is not possible at present. What is revealed is a network of areas, spread out yet interlaced, which, in co-operation with the higher brain, appear to control emotional behaviour by a delicate system of chemical balances. The action of drugs on this 'limbic system', as it is called, is like a shot-gun rather than a target rifle. How could they be otherwise when this core of the brain is as difficult to investigate as the depths of the sea?

Perhaps it is the enormous cortex, out of balance with this old brain, that makes reason push ahead with plans which seem desirable despite the alarm reactions of our emotions. In other words, when rational action is blocked, rationality becomes fierce with emotion in defence of its motives. This continual conflict between old and new brain is a potent source of human ills and perhaps of imagination and invention—madness to great wits is near allied. Only by recognizing this 'fierce disrationality for what it can do to us will we be able to come to grips with it, and therein lies our hope' (A. B. Gilgan).

To work this fantastic machine requires as an energy source a teaspoonful of sugar an hour, and oxygen in plenty to help extract the energy from the sugar. The power derived is equivalent to about 25 watts of electricity, enough to light a dim bulb. Yet this drives the brain—of a Beethoven or a Hitler.

None of these crude facts account for the subtlety, range, and delightful irrationality of human beings. Indeed our view of the brain is like an early map. A few rivers, valleys, and hills marked, and a vague coastline. Perhaps the next century will mark the contours and give the map references to love, pleasure, and kindliness. Perhaps not, for biologists engaged in brain research admit they face one of the most technically intractable problems to confront man.

13: How Chromosomes Affect the Brain and Intelligence

CHROMOSOMES can be seen down the microscope, brains can be dissected and their cells examined, and behaviour can be observed. We can ask the question how chromosomes and their packages of genes might affect the brain and so in turn human intelligence and character. Such evidence is decisive in helping us to determine whether what we are born with affects personality, for the brain is the seat of personality. To be sure, the evidence comes from people with gross chromosome imbalance of the sex and non-sex chromosomes (see Chapters 1, 2, and 5) but the study of the abnormal is one of the keys to the understanding of the normal. The gross abnormalities of brain form and function are, of course, large oscillations on the graph of human variation, and normal sets of chromosomes, each carrying their own genetic pedigree, will give but smaller and as yet unseen variations between brain and brain in, for instance, the number of neurones (nerve cells) and the complexities of the neurone network and in the level of the threshold at the nerve synapses.

OUR FOUR-PART BRAIN

Before we can talk about how chromosome errors affect the brain, it is necessary to mention the structure and function of its parts, although more will be said about the brain in later Chapters. Brain surgeons, when looking at a living brain, say that its strangest quality is its stillness, giving no hint of genius, dullness, criminality, or what you will. Its shape and appearance give no clue to its function, but the fact that nerves run to it from all over the body and away from it shows how the brain has overall control. It is not just one structure but basically it consists of four parts, all closely enmeshed by nerves (see Figures 10 and 13): a double structure, the cortex; the 'old' brain—old in the evolutionary sense; the cerebellum; and the brain stem.

On the outside is the tightly folded, grey matter of the cortex, folded to pack an enormous surface area into the skull. More than half the area of the cortex lies tucked into folds. The cortex is really a double structure split down the middle, but the halves (hemispheres) are joined by rich nerve connections. Some parts of the cortex have a specific job to do. Reports sent to it by each of the five senses are received and analysed by the sensory cortex, complex motor patterns (motor = physical movements) are organized in the motor cortex. If any of these areas is damaged then the corresponding ability is impaired. An accident to the back of the head can damage the part of the cortex where the visual centre lies and may cause total or partial blindness, and so on. In man, one hemisphere of the cortex, the left, is detailed to handle the problem of speech, but speech is more than one thing. Spoken (motor) speech is located in a small well-defined area, but other speech areas—for writing and for understanding spoken and written speech—are not so clearly defined and may vary in position from brain to brain.

If we have two cerebral hemispheres each with a great deal of localized performance, with inputs of general body sensation and vision channelled into one or other side and with movement similarly dependent on one or the other side of the motor cortex, why do we experience 'mental *unity*' and not a jumble of experience arising from the scattered activity of the various areas of the cortex?—undoubtedly, because of the rich nerve connections (commissures) between the two hemispheres. Very recent experiments with cats and monkeys show that information which enters one hemisphere from *one* eye is communicated to the other hemisphere and laid down as a memory trace.

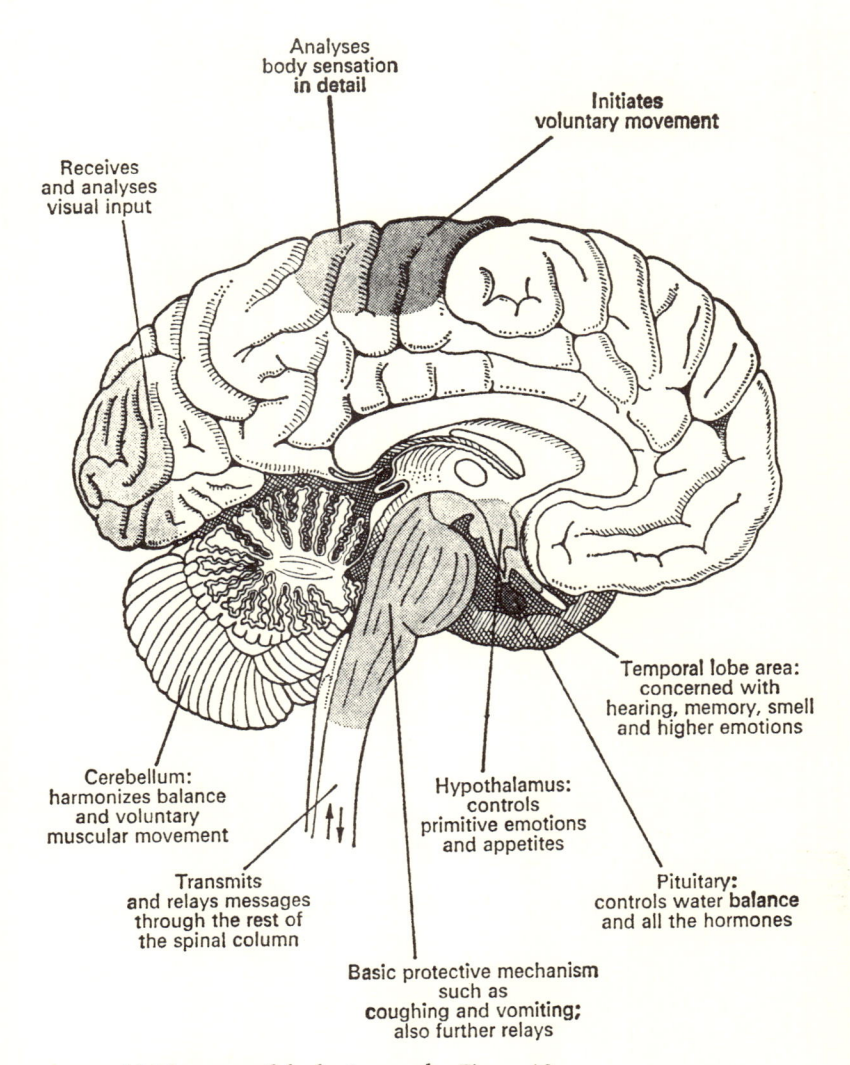

FIGURE 10 The areas of the brain, see also Figure 13.

By contrast, after the brain is split by severing the commissures there is no transfer of what is learned through the eye from one to the other side, and each half of the brain can be trained to give totally opposed responses to stimuli. The brain commissures, then, are essentially concerned with transfer of information between the two halves of the brain so that they can share in memory and learning.

Despite our present knowledge the cortex is still like a map of Africa in 1850, two-thirds uncharted, with only vaguely identifiable jobs to do and only the gross landmarks known. These uncharted, or 'uncommitted areas' of the cortex may well be the main thinking, interpretive areas of the brain (see Figure 11). In a real brain, as opposed to a diagram, the sensory, motor, and thinking parts of the cortex are not neatly arranged in order but are intermingled, though in some order. Imagine a shirt hung out on a line to dry and the same article screwed up in a ball in the washing machine; the parts of the shirt are still in the same order, as are the parts of the cortex. This 'new' brain—new in the evolutionary sense—developed from the 'old' smell brain 'like a splendid flower from a knob of bud'.

The 'old brain' lies deep in the middle of the skull overlain by the cortex. This area includes the thalamus, which is a relay station for sensory messages to the cerebral cortex. There is evidence too that it adds emotional drive to cortical activities that otherwise might be colourless. A sunset and a rainbow give delight, but the delight is based on reason; war can make us feel despondent or angry, but reason, and therefore cortical activity, is mixed up with emotion. Below the thalamus is the hypothalamus pressed against the floor of the skull and surrounded by nerves, blood vessels, and glands. It is the supreme regulator of the internal organs of the body, like stomach and intestines, through the autonomic system, already mentioned in Chapter 11, and also the pituitary gland which lies below it. But again there is a link between the new and old brain as in the thalamus. If you worry in your cortex, the hypothalamus knows and sets the autonomic system on a twenty-four hour alert—hence stomach ulcers, high blood pressure, and spastic colon. We shall have more to say about the hypothalamus later in Chapter 17, but it contains numerous 'centres' for regulating such activities as eating, drinking, sleeping, waking, and sexual rhythms. For most of these activities there are pairs of centres which seem to work, in conjunction with other parts of the old brain, in opposition: one increasing the action, the other reducing it.

The 'smell brain' is another part of the old brain, but in man it is but a vestige of that found in lower vertebrates because of the practical value of smell to them.

As has already been said, the great convoluted cortex in man is derived from the smell brain, but sight and hearing and general sensation have taken over, leaving but a stub for smell. And even this is used for other purposes, connected with emotion. The smell brain is better called the limbic system because it contains centres other than for smell. One of these concerned with memory will be discussed in Chapter 16.

The other two main parts of the brain are the cerebellum and the brain-stem. The cerebellum, or little brain, lies below the back of the cortex and is second only to the cortex in size. It is corrugated by deep side-to-side ridges which increase its surface area. It is really an enormous computer to organize body movements. Perhaps a better description of it is an 'automatic pilot'. It takes over actions merely sketched out by our conscious minds and arranges complex muscle teamwork to bring these actions about, so that we can ride a bicycle while using the cortex to solve an equation or plan the next meal. If the cerebellum goes wrong the cortex itself must carry out the detailed arrangements of the muscles. Actions can be carried out, but they are slow, unsure, tiring, and inaccurate.

Finally, there is the bulb, the part of the brain-stem which continues the spinal cord into the skull. In it are centres for the control of life and death functions like breathing and the control of the tone of blood vessels, also mechanisms which pick out the new and interesting, while suppressing the less important, at a particular moment in time. We pick out our name in the hubbub of a party no matter how fascinating our partner. 'This concerns you', the centres in the brain-stem say, and our attention is alerted. Similarly a mother can sleep through noise yet wake to her baby's cry, and a wounded soldier in action can still carry on because the signals of the alerting centres are hushed by the cortex. These 'alerting' centres in the brain-stem are technically called the 'reticular formation'. They do not, in man, lead to intelligent control alone but only in conjunction with cortical activity. This brings us to a final point about the brain which needs stressing although it has been brought out above. It is that there are probably no particular areas concerned with higher mental processes like problem-solving or creative thought. These seem to be carried on throughout the brain whenever and wherever necessary.

CHROMOSOME IMBALANCE, IQ, AND BEHAVIOUR

The ways in which chromosomes exert their affect on the brain, behaviour, and IQ, are numerous. Genetic imbalance is at the root of the disturbances which often adversely affect brain function and gross brain form. At present there are four main groups of imbalance which we need to know about but new staining techniques will mean that a number of new chromosome abnormalities are likely to be identified in the next few years.

1. The effects of extra *whole*, non-sex chromosomes (autosomes) which can cause such conditions as mongolism.
2. The effects of the deletion of a tiny part of an autosome or duplication of a part, or a switch round or inversion.
3. The effects of a mixture, or mosaic, of normal cells containing normal chromosome numbers with cells containing an abnormal number of chromosomes. Such mixtures can be found of cells containing abnormal numbers of sex chromosomes or of autosomes with normal cells.
4. The effect of sex chromosome imbalance.

These four groups can be broken down into a clinical classification which is helpful: abnormalities of autosomes where patients have severe mental and physical handicaps, and abnormalities of sex chromosomes which quite often give but mild mental defect, plus abnormalities of the reproductive system.

The admixture of normal and abnormal cells as in 3 above often dilutes the effect of the abnormal chromosomes to give higher IQs and less brain damage than 1 and 2.

To find out which chromosome is the culprit in causing the imbalance we need to look at a diagram (see Figure 1) of the twenty-four different kinds of chromosomes in a male cell taken from a normal cell with forty-six chromosomes. The chromosome dimensions and shapes are average in this chart and are based on many observations. Purely for convenience in reference, numbers are given to the chromosomes, grouping them according to the relative length of arms and total size. The nipped-in part of a chromosome is called the centromere and to this the arms are joined. Pair one have arms of equal length, pair five have short and long arms. Most of us, if male, have normal male chromosome sets and likewise, if female, normal female sets, but it has been estimated that about four pregnancies in every hundred carry a chromosome change that may be responsible for abnormal development in 90 per cent of cases. Luckily 90 per cent of these abnormal conceptions abort naturally but the rest survive at birth as abnormal infants. About half the abnormalities are caused by autosome mistakes.

THE EXTRA TWENTY-ONE AND DELETION OF NUMBER FIVE

The possession of an extra tiny autosome, number twenty-one, in every cell of the body causes the distressing and common condition called mongolism. This means that each cell has forty-seven chromosomes, twenty-two normal pairs but twenty-one is in triplicate, not a normal pair. Technically this is called 'trisomy-21'. Actually 'mongolism' is not a good term because there is no relation between the appearance of the abnormality and members of the Mongolian race, although superficially they look like them in the fold of the eyelid. The chromosome imbalance is responsible for the small, light brain, only about 1000 grammes and seldom over 1200 grammes compared with 1300 and 1400 grammes for a normal child at the age of twelve. The frontal lobes of the cortex, the so-called 'silent areas', and particularly the brain-stem and cerebellum, are disproportionately small. There are relatively few 'grey-matter' cells in the third layer of the cortex. This third of the six recognizable layers of the cortex contains neurones connected to incoming sensory nerves and also to nerve cells in other parts of the cortex. The mean IQ of such mongols is about 30–35. Besides being of low IQ, mongols pass their developmental milestones late, not only in sitting, standing, and walking, but in habit training and speech development. The chromosome abnormality thus carries its own distinctive personality pedigree—rather stereotyped. As babies they are good; as infants playful; as children clownish; as adolescents stubborn but amicable; as adults friendly and naïve but not foolish. Unexcitable, content, relaxed, cheerful, affectionate, sociable, mild, and open, are adjectives applicable to mongols. Indeed they are lovable.

The power wielded by the chromosomes is nowhere better demonstrated than in mongol mosaics with normal cells as well as those with trisomy-21. The effect is a dilution of the typical mongol. Instead of an IQ of about 30, mosaics average about 70.

A tiny but rare deletion from the short 'arm' of chromosome pair four or five pushes IQ down disastrously to below 30. The brain here is probably always small, but little is known about the changes in brain anatomy due to this defect. Most of the affected children, however, are short and underweight and have a distinctive cry like a distressed kitten.

Rare deletions from other chromosomes provide their own distinctive pedigree. Deletion of the whole of the short arm of number eighteen, or the whole of its long arm, produce brain damage and mental defect, while a deletion from the long arm of number twenty-one, the chromosome which in triplicate causes mongolism, causes 'anti-mongolism' to give certain features opposite to those of a typical mongol—anti-mongoloid slant of eyes, large and low-set ears.

SEX CHROMOSOME MISTAKES

In contrast to the autosomal, or non-sex chromosome errors, mistakes involving sex chromosomes do not always seem to produce severe mental and physical abnormalities. Indeed the disabilities of people with errors of the sex chromosomes are often solely confined to the reproductive system and many people with such errors are useful members of society. It looks as if man is able to tolerate considerable imbalance of the sex chromosomes without the disastrous effect of autosome imbalance. For example, several men have been found with as many as four X chromosomes and a Y ($44 + XXXXY = 49$) instead of $44 + XY = 46$ in a normal male, while a not very abnormal woman has been discovered with five X chromosomes instead of the normal two.

There are four kinds of sex-chromosome error: the XXY group; the XO group; the extra X group, and the extra Y group. Like the autosome errors, each group has its own distinctive psychological and behavioural 'finger-print'.

An interesting point about 'trisomy' of the sex chromosomes is that with the addition of every X chromosome there seems to be a slight downward shift of IQ. Men with cells containing $44 + XXY$ vary in IQ between 67 and 82, those with $44 + XXXY$ between 55 and 64. Likewise women with $44 + XXX$, $44 + XXXX$, and $44 + XXXXX$ show increasing mental defect with each extra X. Mosaics such as

44 + XY (normal)/44 + XXY appear in sub-normality hospitals five or six times more frequently than among newborn males. These chromosome mosaics seem to have a higher IQ than non-mosaic but chromosomally abnormal people, but the admixture of normal and abnormal cells in the brain might have an adverse effect on its more subtle functions and on the capacity for social adjustment of mosaic individuals, and perhaps this explains their relative frequency in sub-normality hospitals.

While the effect on the brain is not known, it is quite likely that microscopic or sub-microscopic changes do occur with the addition of each extra X chromosome. Perhaps relevant to this argument is the fact that the number of ridges in the finger-prints alters with sex-chromosome errors. A normal female with XX might have a total of of 127 ridges on the fingers, an XXX woman 108, a woman with XO (only one X instead of 2) 167.

For each extra X chromosome the total ridge count is depressed by about thirty or so ridges. If finger-print pattern is changed, why not brain cells? The mechanism of these finger-print pattern distortions might be due in some way to a disturbance of the fluid content of tissues by the chromosome errors: too many Xs might cause cellular enlargement and therefore contraction of the finger-tip; with one too few (XO) Xs, the cells become smaller due to fluid collecting *between* the cells and therefore expansion of the finger-tip. How these chromosome disturbances may affect brain function and how this effect is mediated through the shape of cells and their size, number, and chemical characteristics, is unknown. Men with XXY make-up (the Klinefelter group) are often shy, childish, and timid (the opposite of the extra Y), lack interest, endurance, and originality, are indifferent to sex, and do not seem to be able to form strong ties with people. An extra Y, as we have seen in Chapter 10, results in men of formidable height, 1.9 metres on average compared with 1.7 metres for most males, a record of crime more against property than against people, an early start to a career of crime (thirteen years) and, compared with men in the same hospital, they come from almost crime-free backgrounds. In association with the increased body size of extra Y men, there is a broadening of the finger-print ridges. The extra Y, given a suitable environment, propels some males towards early crime. However, most have a normal IQ, only about one-quarter are sub-normal, and a few cases have been discovered with an IQ as high as 134.

Sometimes women are discovered with one X chromosome missing, to give 44 autosomes + XO = 45 (Turner's syndrome). Physical abnormality includes the fact that they are small, have no true ovaries, and are thus infertile, but on average their (verbal) IQ is normal—97. Curiously, as indicated in Chapter 5, they do badly in tests of 'spacial perception', that is the kind of tests that involve shapes and patterns. Sometimes their performance is so bad that their condition is called 'space-form blindness'. In one instance where XO patients had to do a road map test, they had to indicate on the map where they would make right-left turns on the 'walk', and the number of errors they made was considerably greater than a group of nurses who acted as a control. People with Turner's syndrome are thus at a disadvantage in an IQ test where shapes and patterns predominate because of something quite specific in their inheritance which influences the brain. Significantly, mosaics with cells containing the XO chromosome pattern and cells containing the normal complement have a higher verbal IQ, 110 on average, but whether their spacial perception is better than that of XO individuals is not known.

All these chromosome abnormalities make their own distinctive and cruel mark on brain and behaviour clear indication of the power and precision of heredity. Those of us who are lucky enough to have 'normal' sets of chromosomes also have 'finger-print' individuality through the interaction of unique genotype and unique environment. Fortunately the harmonious balance of genes allows a marvellous range and subtlety of behaviour and thought that gross chromosome errors do not seem to allow. But the abnormalities so described, together with the rest of the evidence in this section, show decisively that human character and behaviour is prescribed within the limits laid down by the character of the fertilized egg. Is this a gloomy conclusion? Perhaps it is, but we must remember that man has profited by understanding the laws governing the physical and biological world. Similarly, by understanding his own nature and its laws man might be able to guide his own life and future better. We know too that each one of us, except identical twins, is genetically unique. In other words, each one of us has a brain that is individual and differs from every other living brain. We shall see later that the 'wiring' of the brain is partially under environmental control, and according to the richness of early childhood experiences (up to about two years), so the brain can alter. Such knowledge is a charter of individuality: to enhance the indelible character and personality of each one of us for the betterment of ourselves and mankind. The basis of this individuality lies in the awesome complexity of the brain, the study of which, let us admit, is still at an extremely primitive stage and gives but a dim and shadowy picture of the way it might work.

14: The Brain as a Machine

J. B. S. HALDANE, in a talk delivered in 1963 on biological possibilities in the next 10 000 years, said that the exploration of the interior of the human brain would be as dangerous as that of the Antarctic Continent or the depths of the ocean—yet more rewarding. Haldane's time-scale helps to make one or two points about the brain worth stating at the start of this section. The first is, to some extent, a declaration of the present state of ignorance about brain machinery which is understandable because this is extremely complicated, delicate, and hidden. When viewed coldly it does seem remarkable that about three pounds of sloppy jelly could be at the root of the complex, subtle and sensitively co-ordinated actions of which man is capable. Although, to be sure, something is understood about how the brain works, judging from the volume of the reports of modern discoveries and spectacular treatments for mental illness, a sober view is taken by great neurologists like J. Eccles, who declares that our knowledge is at 'an extremely primitive stage' and gives but a 'dim and shadowy' picture of the brain's amazing intricacy of pattern. Just how ignorant we are is underlined by the fact that nothing in the whole of brain research undertakes to explain how information from the eye when relayed to, and activating, the cortex, gives a three-dimensional coloured picture of the outside world. Even more wonderful is that in a dark and silent room, memories of the past can well up in our minds as fresh and clear as direct experience through eyes and ears.

Perhaps rather more to the point than asking the question, 'What do we know about how the brain works?' is 'What are we finding out?' because views on the brain's working seem to change so frequently. In short, no clear theoretical structure, no firm knowledge has yet emerged about the brain's working. But useful discoveries are coming in from many different sciences. The task of neurobiologists seems to be that of trying to fit together an enormous jigsaw puzzle.

Second, an apology. Any description of the brain tends to end up as a catalogue describing the structure and function of this or that part. Such disjointed treatment is perhaps inevitable with the fragmentation of knowledge, but it allows no real insight into the function of the brain as a whole. The sketch of the brain in Chapter 13 tried to indicate this 'all-of-a-piece' functioning. This section of the book falls to some extent into the trap of fragmentation, but the reader should remember what has been stated above.

SOME FACTS AND FIGURES ABOUT THE BRAIN

The brain of a normal adult weights on average about 1300 grammes and accounts for about 2.0 per cent of body weight. Yet with the destruction of the brain all the rest of the body loses its point, whereas a partially destroyed body with an intact brain retains its human quality. Brain weight is not necessarily correlated with ability but the cut-off seems to come in adults below 1200 grammes. Certainly less than 1000 grammes is incompatible with normality. (See Chapter 13.) At birth a brain weighs about 325–350 grammes, by the age of one, 800–900 grammes, and by six or seven it has reached adult weight. A point has already been made in Chapter 7 about the brain's avid need for protein and the link between poor nutrition and IQ. At birth, all the deep fissures which give the cortex its mighty surface area are practically complete.

Until recently it has been assumed that no new neurones are formed in the brain after birth and that a baby has its full complement. Altman has shown recently in rats that this is not so and the same may be true in man. Neurones do not divide, but new small neurones are formed after birth and these migrate between the larger neurones of the cortex and attach their axons to them as they pass. Not only this, but the neurones of the newborn are very different from those in the adult. During the first two years of life the dendrites (see below) sprout many branches, and such proliferation is said to be under environmental control. All these findings emphasize the uniqueness of the brain, because the 'wiring' pattern must differ from individual to individual according to heredity and environmental influences. Rats raised in a stimulating environment, compared with rats brought up in a poor one, have heavier brains due entirely to the growth of the cortex. Indeed, in rats different cortical areas can be selectively developed by altering different aspects of the environment.

The cortex, the human thinking cap, forms a layer of grey matter about three millimetres thick spread over the whole surface of each half (hemisphere) of the brain. If the cortex could be unfolded and spread out, it would form a sheet of about 2000 square centimetres. Beneath the grey matter of the cortex, grey because it contains the cell bodies of the neurones which we shall return to in a moment, the bulk of the tissue is white matter—the pathways of fibres (axons) of the neurones.

Rough estimates of the number of neurones in the cortex, made by counting the cells in small slices of tissue and multiplying up, suggest there may be ten thousand million neurones packed into it. It is not just the master level of the human brain, the cortex, that contains neurones; the cerebellum contains fantastic numbers, and other regions contribute too. Besides the neurones there are supporting and other types of cell, so that the brain might contain in all a hundred thousand million cells packed into some three pounds (about 1300 grammes). In a way the vast numbers of brain cells is as paralysing to our comprehension as the vastness of space, and has prevented neurobiologists from making much progress.

The fine structure of the brain, the neurone, needs mention. Figure 8(b) shows the shape of this immensely complicated cell. From the main cell body, which gives the cortex its grey look, sprouts a single axon, essentially a 'wire' covered with white insulation, the white matter of the brain. This carries the nerve impulse, not as a continual stream of electricity like water through a pipe, but as a series of impulses like bullets from a machine gun. From the main cell body other branches extend. These are dendrites, which may be few or many, and simply or profusely branched. They receive messages from sense organs or, in most cases, from other neurones. These messages may be of different kinds, and a neurone combines or 'sums' the impulses it picks up through its dendrites. On the basis of this information, a neurone produces or does not produce a pulse of electricity. At the end of the axon, the pulse of electricity causes a flood of chemicals to be released (transmitter substances) which cross the gap between neurones to stimulate or inhibit the next neurone in the chain. The sub-microscopic gap between two neurones is the synapse. (See Chapter 11.) The synapse acts as a kind of valve, forcing impulses to flow in one direction so that messages do not meet head-on. Such a system is described as 'polarized'; that is, it has two distinct ends going from A to B with all kinds of adjustments made by the synapses going on in between. The synapse seems to act as a super-filter, sorting important messages from trivial ones, and merges simple messages into complex blends. It might, too, provide the basis for memory, but more of this later in Chapter 16.

At this point it is worth digressing to mention an ignorance we are overcoming. When light falls on the retina of the eye, messages of an electrical nature are sent along neurones to the parts of the brain known to deal with sight. What is sent to the brain is not an accurate copy of the picture but a simplified electrical and chemical code which somehow the brain turns into a picture. The eye is not simply a television camera reporting all that takes place. Rather it is a selective filter letting through only what seems important. We shall come back to this in Chapter 15.

Returning to the neurone network, each neurone is linked by synapses to hundreds of other neurones, in some cases, it seems, to as many as a quarter of a million. When a nerve pathway is used a good deal, as we have seen in Chapter 11, the 'threshold' of the synapse falls so that it operates more and more readily. An impulse discharged from one neurone causes a momentary activation (or inhibition) in the synapses which each neurone forms with other neurones, but for an effective spread of activity each neurone must receive synaptic activation from hundreds or thousands of neurones and itself transmit to thousands of others. A 'wave-front' of activity is thus started which might sweep over at least 100 000 neurones in a second, comprising, as J. Eccles states, 'a kind of multi-lane traffic in hundreds of neuronal channels'. An advancing wave would also branch at intervals, often abortively, and join with other waves to give a complex and fleeting pattern. During unconsciousness—due to concussion—or in deep sleep, there is a very low level of neuronal activity. By contrast the electrical activity of the wide-awake brain shows that millions of neurones are involved in intense and varied activity. Not all the neurones would be discharging an impulse; many would be 'critically poised' for discharge and it has been postulated that consciousness is dependent upon the existence of a large enough number of critically poised neurones superimposed against a high level of background activity. Only in such conditions might 'willing and perceiving' be possible.

The impulse down a nerve depends not only on electrical but also on chemical energy. Where does this energy to drive the brain come from, and what roughly are the quantities of energy used by the brain? The raw materials for the driving power of thought, vision, co-ordination of the body, and all the other activities centred in the brain come from glucose and oxygen, and the brain's need for these is continuous and avid. Glucose at the rate of one teaspoonful each hour is the basic energy food of the brain, and oxygen is necessary to make the sugar yield up its energy. Oxygen is consumed by the whole brain at about forty-six cubic centimetres per hour, which is about a quarter of the total oxygen used by the resting body, but individual parts of the brain, for example the visual cortex, when working use oxygen at a greater rate than other parts. As we have seen the breakdown of this amount of glucose yields only enough power to feed a dim light bulb, yet this little power supply enabled Mozart to write a concerto, Rembrandt to paint a masterpiece, and enables each one of us to do our daily work. How different from the millions of watts needed to keep a computer going to perform its relatively restricted operation. Eating more sugar or breathing more oxygen will *not* better the quality of the thinking any more than putting more current through a television set will better its performance. Taking 'normal' consumption of oxygen as 100 per cent, the sleeping brain uses much the same amount of oxygen (97 per cent) as the wide-awake brain doing mental arithmetic (102 per cent) though there is a redistribution of oxygen and energy use in the sleeping brain. Severe mental illness like schizophrenia does not alter the oxygen consumption of the brain, so it takes just as much oxygen (and therefore energy) to think a queer thought as a normal one. An alcoholic coma reduces oxygen consumption by half the normal amount.

The study of how the brain works is at a primitive level, not only because of its delicacy and complexity but also for the less obvious reason that a language for describing the activities and 'what it is for' has not been easy to find.

SCREWS AND COMPUTERS

Fortunately, with the development of computer engineering a language has been found and analogies made which have enabled brain function to be studied fruitfully. In comparing brain with machine, however, we should not forget that in the nineteenth century the mind was thought by some to be a highly complicated machine based on sound engineering principles. Failures of mental function are still described in the language of the mechanic—stress, strain, a loose screw, nervous tension, etc., yet of course there are no wires in the brain, nothing that reaches breaking point and snaps. Similarities between brains and computers have created impetus in research, and computer language has crept into our everyday speech. We are switched on or off or recharging our batteries, but again there are no switches to throw. Our descendants may smile at the terms we use now to explain brain function (they will have their own brand) but to be fair computers have provided a useful analogy with the brain. Just as important is the fact that writing programmes for computers has made people consider closely how reasoning takes place.

Computers can now be built for storing, transmitting, and manipulating information. Some machines with a simple built-in programme can even be made with a capacity of behaving in ways unpredicted by their designers because an element of random choice can be built into them so that the final behaviour cannot be forecast with certainty. They appear to have a will of their own. The simpler the built-in programme, the fewer are the stereotyped responses and the more human is the computer. The most useful of these machines from the point of view of brain analogies are those whose rules of behaviour are adjusted according to the degree of success they achieve by trial and error, and they do this by adjusting the relative probabilities of different patterns of activity. In other words the computer's rules are modified statistically as a result of its own trials and errors. This is a bit like the human brain, which has an enormous capacity for learning new responses and modifying them in the light of experience.

Machines like the ones described have demonstrated that networks of a few thousand on/off switches (transistors) wired together are what is needed to produce certain 'brain-like' activities: storing information in code; the ability to calculate the probable outcome of an action by comparing new information against encoded past experience. In other words they are 'intelligent', 'purposeful', and so on. These self-guiding machines with a kind of artificial intelligence are teaching us that the brain does not simply decide on a course of action and then go through with it. The brain, like the calculating mechanism, is a self-guided missile, must continually modify performance and revise a plan in accordance with a mass of data from sense organs and from past experiences recorded in its storage devices. The brain monitors an action about five times a second so the issue of fresh plans every fifth of a second, based on the latest sense data and matched with relevant past experiences, allows a plan to be stopped or modified if anything new and urgent crops up. The approach of the hand to a cup of tea is observed and error corrected, but the sight of lipstick on the cup causes us to revise our plan—all rather obvious when explained to the layman.

Nevertheless, despite much progress mechanical imitations which mirror man's brain are a long way off. Like the digital computer, the brain consists of a network of switches, the neurones, but there are ten thousand million neurones in the cortex alone, and the biggest computer in the world has less than a million transistors. Each switch in a computer usually connects to two or three other switches, but as we have seen each neurone in the brain is linked to hundreds or even thousands of other neurones. To imitate a single neurone would take a fair-sized computer.

In a computer a transistor receives electrical impulses telling it which way to switch. The neurone also selects messages and decides whether to conduct them nor not. Like the transistor it gives a crude yes/no choice. It is the synapse, as we have seen in Chapter 11, that makes possible the subtlety and complexity of human action and thought, and here the chemistry of the synapse is linked with the electrical impulse of the axon. Indeed it is because of the synapse that the neurone network outshines the transistorized computer in many ways.

The synapse can sift and select important messages, and the flash-point, or 'threshold' at which a neurone discharges its impulse, depends on the height of the threshold in the synapse over which an impulse must 'climb'. Use of a nerve pathway lowers the threshold and so it operates more and more readily, but it will drift back to its normal height when disused. Nevertheless our attempts to imitate the brain with small and crude computers do produce brain-like behaviour and this is an important step in understanding what the brain does.

Perhaps the most useful function of a machine like a computer is its use to test some hypothesis about brain function, when it serves as a tool for research. D. M. MacKay suggests that we can think of a mechanical brain model as a kind of template which we construct on some hypothetical principle and then hold it up against the real thing (the brain) in order that the discrepancies between the two may yield fresh information. This in turn should enable us to modify the template in some respect, after which a fresh comparison may yield further information, and so on. The model, as it were, 'subtracts out' at each stage which we think we understand, so that what is not yet understood is exposed more clearly.

Is the brain, then, just an astronomically complicated machine which generates the mind, something that in the next thousand years MacKay's machines will imitate perfectly? Is thought and consciousness the result of a flow of electricity (and changes in synapse chemistry) through the webs of neurones? Or is there some mystery to the brain that will never be plumbed? Many scientists would say that the human brain-plus-body is no more than a cybernetic mechanism, though they would stress that our ignorance of brain function far exceeds our knowledge.

In this chapter we have seen that the brain is a little like a digital computer because it consists of neurones triggering one another to fire through nerve webs. But the electrical/digital analogy should not be pressed too far or make us forget that the brain is a chemical instrument too. If alcohol is poured into a computer it will not affect its rate of operation, but the effect of alcohol on the brain is profound, as we know. If carbon dioxide, a by-product of sugar breakdown and energy release, is not shifted away from the brain by the circulation, the brain is soon, as it were, choked by its own smoke. We know that electricity and chemistry are closely integrated in nerve activity and we shall deal more specifically with chemistry in Chapter 17.

15: What the Eye Speaks to the Brain

IN this chapter we shall consider some further clues on how the brain works; more specifically, the way in which the cortex might sort out the light waves that reach our eyes into patterns we recognize.

Dr Richard Gregory, the Cambridge psychologist, once said 'if we hadn't grown eyes we wouldn't have needed a brain—to my mind we got this capacity to think almost as a by-product. As you'd expect the brain's a pretty rough old job.' The reader must decide for himself how rough the job is, but the study of what the eye tells the brain—the problem of vision—gives an insight into the complexity and difficulty of discovering what the brain does.

This is no place to go into the structure of the eye except to say that the retina, the paper-thin lining of the eyeball, is the true sense organ that detects the light. The retina is an outgrowth of the brain and is connected to it by another outgrowth from the brain, the optic 'nerve'. The neurones which make up part of the retina are connected to special cells, the rods and cones, which are light energy detectors. The rod is the most common light receptor in the human retina and estimates put the number of rods in a pair of human eyes at about 130 million. Cones number about 7 million. Rods and cones are in connection with the neurones, and there are about 450 000 of these in each optic nerve. A great number of rods, around 150, are connected to one neurone. Rods are useful in dim light; they have what is known as a 'low threshold' of vision. For reports of vision to occur in the brain, between 5 and 14 light quanta must fall upon an area containing 500 rods. Statistical calculations based upon these results show that between 5 and 14 rods must be stimulated for the brain to know of the stimulus. It follows from this that sometimes a very faint light near the 5 to 14 threshold is seen and sometimes not. Cones were once thought to have their own private 'line' to the brain, but since there are about 8 cones in the retina for every neurone in the optic tract, sharing must occur. The cones, which are found mainly packed in a special circular area of the retina (the fovea), work best in high light intensities; thus they have a high threshold of vision. Cones are responsible for fine discrimination.

When light energy falls on the rods and cones they change chemically; the greater the light intensity the greater the chemical change and vice versa. So when a pattern of light and shade falls on the retina focused by the cornea and lens, a corresponding pattern of chemical events is formed on the retina. But the pattern is not fixed. Like the continually changing world around us, the chemistry of the retina is in rapid flux.

Every change in the chemical patterns formed on the retina triggers electrical nerve impulses which are relayed to the brain, to its 'visual cortex' which is, surprisingly, not behind the eyes but at the back of the head. In man about one-tenth of the cortex is given over to sight, a far greater proportion than to any other sense, smell for instance, since most information in man comes in through the eyes. The precious visual cortex is well protected, most of it being tucked away into a deep fold. The cortex consists of two hemispheres, and the 'screen', which receives the electrical impulses from the retina through the neurones, is in two halves, yet we see one object.

The route taken by the optic nerve to the visual cortex is quite complicated. From the back of each eyeball the optic nerves pass into the skull and form an X-shaped cross-over, the chiasma. Here the nerves are re-arranged so that nerve fibres from the left half of each retina pass to the left side of the brain and vice versa. The upshot is that half the fibres cross over and half stay on their own side. The reshuffled fibres leave the chiasma to form the optic tracts and these lead to the thalamus, to which we shall return later. It is thought that the thalamus, a complex structure, may somehow organize and modify the sensory messages and give them emotional colour before they are relayed to the visual cortex by a sheet of nerve fibres. The functioning of the thalamus appears to be precise, since distinct thalamic cells are aroused when a certain part of the retina is stimulated and these cells in turn activate a small portion of the visual cortex.

Every tiny group of retinal neurones sends a nerve axon into the optic nerve and the relative position of these fibres in the nerve, in the thalamus, and in the visual cortex, is preserved. Hence stimulation of a set of rods and cones in, say, the form of a square produces an approximately square area of activity in the cortex. If the retina is illuminated by a faint, narrow beam of light, electrodes on the cortex record localized activity and the point of maximum activity moves with the beam, a shift of one degree producing a movement of one millimetre in the cat, more in the monkey.

THE TIME-LAG BEFORE A CONSCIOUS EXPERIENCE

Some interesting work by B. Libet has shown how long it takes for the elaboration by the cortex of a conscious experience like seeing a flash of light or experiencing a feeling of deep pressure, 'like a wave moving about under the skin'. He (and others) applied to the exposed brain cortex of conscious people very weak trains of brief electric pulses at frequencies of thirty to sixty a second; the object of this was to discover the length of stimulus that just sufficed to give them a conscious experience. Surprisingly, at least half to one second of repeated stimulation is required to produce a feeling of pressure, or pain or cold, after the initial arrival of impulses at the cortex. With the visual system, at least one-fifth of a second of cortical activity is necessary before a weak flash of light can be detected. The 'incubation' period for simple sensory developments at the cortex is, therefore, very long indeed. Why? The time for an impulse to get from one neurone to another is no more than one-thousandth of a second; hence there could be a serial relay of as many as 200 synaptic linkages between nerve cells, with a fraction of a second's pause at each synapse before the impulse continues and a conscious experience arises 'in the mind's eye'. Compare this relatively small activity with everyday experiences. Looking through the window as I write, with thunder in the air and lightning flashes raking the sky, these stimuli received by their appropriate sense organs must cause millions of neurones in the cortex to form and reform new and highly complex patterns of activity. These must occupy not only the cortical cells activated originally by sensory experiences but others independent of them, involving as Eccles writes, 'comparison, value, judgement, correlations with remembered experiences, aesthetic evaluations' and so on. In short there must be quite fantastic complexities of neuronal patterns operating in the cortex at any one time.

If the lag between sensory stimulation and cortical activity is great for the development of a conscious experience, it is not so for motor reactions, where a reaction time as low as 0.05 of a second can be demonstrated. Perhaps the quick motor reactions we make in response to everyday life, like driving a car, may be mediated by processes at sub-conscious levels, and perhaps only after our actions do we become consciously aware of the sensation and response—or we might remain unaware of it altogether.

THE EYES AND BRAINS OF FROGS AND CATS— CLUES TO HUMAN EYES

A baffling question now arises: how do we sort out, from the light waves which continually bombard us, objects that we can recognize? Clues to this come from some brilliant work on the frog's eye by J. Y. Lettvin and others, and by D. H. Hubel and T. N. Wiesel on the cat's eye.

Because it is relevant to the findings, it is important to note briefly basic differences between the frog and cat as far as behaviour is concerned. A frog will starve to death surrounded by plenty of food if the food is not moving and if it is not the right size. He will leap to capture any object the size of a worm or insect if it *moves*. As far as escape from enemies is concerned his actions are simple; he will leap to anything darker than the moving object that might be an enemy, which could be water or land, for he can live happily in both. We might say that the frog's visual apparatus is specialized to recognize a limited number of stereotyped situations. The cat is, as we all know, versatile and intelligent and his eyes are sharp as needles. As for the brains of both animals, the frog has no visual cortex, a fact which throws more emphasis on the retina as a sorting mechanism, while the cat, like monkeys and like man, has a well-developed visual cortex.

The methods involved in finding out what the eye tells the brain are extremely complicated and this is no place for detail, but they involve placing tiny electrodes in single nerve fibres in the optic nerve of frogs and in the visual cortex of lightly anaesthetized cats. The cats were shown patterns of light on a screen, the frogs various shapes held by magnets to an aluminium hemisphere. While the animals were shown the patterns electrodes were placed in brain cell after brain cell in the cat, and likewise in nerve fibres in the frog, until one was found which discharged an electrical impulse when the stimulus was presented.

What happens in the frog? Briefly, the electrical activity of the nerve fibres shows that the retina detects four kinds of stimulus about which it informs the brain:

1. local sharp edges and contrasts;
2. sharply curving edges (hence small blobs);
3. movement of edges;
4. dimming—response to a sudden reduction of light.

The nerve conduction to the brain from the dimming detectors is fast, ten metres per second compared with twenty to fifty centimetres per second from (1) and (2). In the frog's retina there are thirty times as many of the first two types of detectors as of the latter two.

Each group of detector fibres maps the whole retina, not just one spot, and ends up in a single sheet of endings in the frog's brain, and there are four such sheets in the brain. What do all these mechanisms amount to in the life of a frog? First, they mean that the retina 'speaks to the brain in a language already highly organized and interpreted, instead of transmitting some more or less accurate copy of the distribution of light on the receptors'. Second, the visual mechanisms appear to be well suited to a life of food-getting and escape. The sharply curved dark edges must represent insects, and since these 'bug detectors' only work on moving objects, the frog is not distracted by spotty patterns in the background which do not move. One interesting detail here is that a large colour photograph of flowers and grass waved in front of a frog, gains no response. If a fly-sized object is perched on the picture and moved over it there is great activity in the bug-detecting nerve fibres, but if the fly is fixed to the picture and the whole moved there is no response.

Now consider the cat. It appears as if the *brain* (not the retina) here looks for:

1. contrasts of light and dark—edges, slits, and bars (like the frog);
2. orientation of these, whether vertical, horizontal, or slanting;
3. again like the frog, direction of movement—a highly ingrained pattern in the cat and the last to vanish with deep anaesthesia.

The regions in the visual cortex that 'look' for these are columnar in shape and about half a millimetre in diameter, and extend from the surface of the cortex to the white matter. These columnar areas might work on their own or in conjunction with others. Perhaps in the visual cortex there are a dozen different types of columnar regions which recognize different angles. Within each columnar area there seems to be a hierarchy of cells, some with simple visual 'fields' (which discharge an electrical impulse when slits, edges, and bars are shone on the retina) and some with complex fields. These might assess the information passed up to them by simple cells, to determine the length of a line or to recognize three lines in a triangle. Cells which fire in response to line length and corners have been found, and there are 'looming' detectors which fire in response to approaching or receding objects. We know that there are other cells in the columnar areas of the cat's cortex which are affected by light touch or bending of hairs or manipulation of joints. Perhaps one day areas will be found that 'fire an impulse on seeing a face shape and which respond to numbers—"two-ness" and "three-ness"'. Indeed, in the monkey, C. G. Gross has found complex cells in the cortex that fire on seeing a paw shape and others that fire on seeing a monkey's face.

Generalizing, we see now that what the eye tells the brain is a code, not precisely of the light patterns that fall on the retina but broken up into information before it ever gets to the brain. We have seen that there are two principal types of visual system. The first is typified in the frog. The individual ganglion cells of this animal are highly specialized in terms of stimulus requirements, and such fundamental variables as edges, colour, contrast, orientation, and directional movement are processed intensively *within* the retina.

The second type of visual system is the one found in cats, monkeys, and possibly man. Here the ganglion cells at the retinal level are concerned only with the simultaneous contrast between the centres and the surrounds of their 'receptive fields' and in some cases with colour information. (The receptive field of a given nerve cell is simply the area of the retina that will cause firing of the cells when stimulated by a small patch of light.) In cats, some cells fired when illuminated at the centre of the field ('on' discharges) and the illumination of the periphery induce 'off' discharges. Others are just the reverse with 'off' discharges in the centre and 'on' discharges at the periphery of the field. The aspects of edge detection, orientation, and directional selectivity, are dealt with only later in the visual cortex in the brain, and there in a most detailed and precise manner.

A MEMORY ALPHABET?

We have mentioned above that neurones have been found that fire in response to shape, and an interesting theory put forward by J. Z. Young relates to this and is concerned with what he calls a 'pre-set alphabet' laid down by heredity, made up of brain cells and capable of keeping an account of events and remembering things in almost unlimited number. Microscopical work has shown that brain cells do differ in shape, and their peculiar character is given to them by the dendrites, the sprouting tendrils which form the receiving gear of the neurones. Young's idea is that the differing shape of the dendrites might form the memory alphabet.

The basic requirement of a dendrite-alphabet is that it is a suitable one to store representations of all the situations an animal is likely to meet in a life time. Evidence from the frog's eye–brain experiments mentioned on page 120 shows that its visual alphabet is simple and allows storage of information likely to be of interest to the frog. It can, as we have seen, recognize blobs that move (bugs); it also jumps towards blue rather than other colours—towards water rather than vegetation.

Our own visual alphabet, if it exists, must be a good deal more complicated. Some evidence for its existence comes from the work of an American infant teacher (R. Kellog) in California and two German scientists (M. Knoll and J. Kugler). The teacher analysed 300 000 drawings and paintings from two-year-old American, Chinese, French, English, and Negro children, and concluded that they are built up from twenty basic 'scribbles': arcs, waves, lines, circles, dots, spirals, poles, and others; and six typical 'diagrams': crosses, squares, circles, triangles, and so on*. The child combines these scribbles and diagrams into more complex forms, so by the age of four he can draw such things as the

*See *Biology and the Social Crisis* (London: Heinemann Educational Books, 1967.)

sun, a house, people, and flowers. The scientists have been investigating luminous patterns seen when the eyes are closed and when the brain is stimulated electrically. Copies of 520 of these 'phosphenes' observed in several hundred people have been collected and analysed into 15 groups, and it is striking that 90 per cent of the phosphenes can be found among the basic scribbles of children. Both have in common arcs, crosses, waves, triangles, poles, spirals, and other patterns. It looks, then, as if a kind of visual alphabet might exist in the cell components of the brain common to all races.

PICTURES IN THE MIND

Somehow this memory alphabet, which of course is only a theoretical concept, might be used as a model against which new information, fed in through the senses, can be matched. Before a decision is reached and action taken, small-scale experiments in the mind, within the bounds of the model, go on in the brain. When a forecast has been reached which does not conflict with any rules of conduct the animal, or man, has learned through experience that action or decision follows. We might alter our behaviour in accordance with changing external conditions, as when we drive a car. Our brain forecasts our position in relation to other moving traffic in our mental picture (for all this is going on in our head) and our life depends on the accuracy of our forecasts.

This notion of a model of the outside world, actually inside our head in the brain, is a difficult one to grasp. We are so familiar with it that we think it is the outside world, and primitive simple people think that when they look out of their eyes and see what is before them, they are performing a positive act. But the interpretations are going on within our heads, as we have seen, by electricity triggering chemistry. What we are really aware of is an imitation world, a tool which we can manipulate in the way that suits us best and so find out how to manipulate the real world which it is supposed to represent.

The picture or model is a highly personal one. A baby has no model, only, perhaps, its inherited brain-alphabet from which it will be able to construct its own model of the world. By staring, feeling, and tasting, it will slowly make a construct of its surroundings (remember here the impoverished world of the down-town white or black mentioned in Chapter 7 and its effect on the brain) and learn how to use its muscles within the framework of its mind-picture. Learning at this stage is slow and inefficient.

That such an inherited brain-alphabet exists is suggested by some experiments by D. C. Fantz on the way in which infants of four days to six months respond to various approximations of the human faces. Infants were tested with three flat objects the size and shape of a head: a painted stylized face in black on a pink background; a face with the features rearranged in a scrambled pattern, and one with a solid patch of black painted at one end equal to an area covered by all the features. The results were about the same at all the age levels. The infants looked at the 'real' face, less often at the scrambled face, and ignored the third pattern.

In another critical experiment, the babies were offered a choice between a solid sphere and a flat circle of the same diameter. The solid sphere was more interesting to the infants from one to six months old. Both the face and the sphere-circle experiment suggest that the brain may be keyed genetically to respond to patterns like a face, and to patterns associated with a solid object which give the infant a basis for perceiving depth.

The newborn infant, then, does not appear to start from scratch in learning to see and to arrange pattern stimuli. This genetically determined response to pattern provides a basic foundation for the vast accumulation of knowledge by experience.

Simultaneously with the building of its model, early in life the dendrites are sprouting new branches to form subtle and complex neuronal patterns, while the small migrating neurones add to the individuality and complexity of the brain's wiring pattern.

PLASTICITY OF THE BRAIN

These last statements about the plasticity of the brain's wiring patterns are supported by experiments with cats and kittens. If one eye of an adult cat is covered with an opaque substance that lets in light, but through which no pattern can be distinguished, there is no change in the behaviour or electrical activity of the brain cells after three months: the cat is normal. Kittens, on the other hand appeared to be totally blind in the eye which was covered for three months. The brain cells did not die, but the nerve synapses appeared to have degenerated through 'lack of experience'. After fifteen months the kittens remained blind.

Another experiment shows, too, that when adult cats are fitted with spectacles on which either horizontal or vertical stripes are drawn, and made to wear them for several months, the cells in the visual cortex responsible for detecting orientation continue to fire normally when the spectacles are removed. So a cat that has worn spectacles with horizontal lines can still detect vertical, and vice versa. It is normal in its response to pattern. It looks as if the wiring patterns of the brain after a certain time lose their plasticity and become fixed. The cells in the visual cortex of kittens which have undergone the same experiment, however, only fire at the patterns they have got used to. The cells will not fire at horizontal or nearly horizontal patterns if they have been fitted with vertically striped glasses. There appears to be no change in cell number and it appears that the wiring patterns have changed, though this seems to be possible only when the cat is young. All this evidence, if the work on cats can be applied to men, has enormous significance for the education of very young children through a stimulating environment.

The massive organization of the brain by early experience is probably why man needs a long and sheltered infancy when learning can take place. Throughout life the brain model we have referred to is enlarged and added to and is highly personal, not only because of unique 'wiring' patterns but also because experiences are *selected*: the brain is no mere passive spectator, but rather a filter. Some things rivet our attention and some things we ignore. Those that focus our attention might well leave a permanent brain record, while those that do not leave none. If the brain was undiscriminating and dealt with all the information supplied by the sense organs, it would be quickly overwhelmed. It seems to be the reticular formation in the brain stem, a core of small nerve cells rich in blood vessels, with profuse connections with most of the rest of the brain and spinal cord, which is concerned with the level of arousal and makes us operate more efficiently. This mechanism might perhaps select the new and interesting and play down the less important, so that, for example, a mother wakes to the sound of her baby's cry while sleeping through the roar of a motor bike.

Even though the brain picture is the product of a machine, it is a highly individual one depending on unique inherited components—the characteristics of the neurones, their individual wiring patterns which seem to depend to some extent on the quality of environmental influence, memory mechanisms which might well differ from individual to individual and culture to culture, and, most important, a great degree of selection. The practical usefulness of this picture or model is that we can use it to guide and plan our actions.

In this chapter we have mentioned memory and we now need to explore this very complex mechanism in more detail. How, for example, is memory organized, and is there a place in the brain where memory is localized?

16: Memory and Learning

IT is commonplace to the layman that two kinds of memory short-term and long-term, exist in the brain, and brain research is confirming this. We use short-term memory to hold a telephone number between directory and dial, furiously repeating it lest we forget it, though if anyone interrupts us we forget it very rapidly. Indeed after a few seconds or minutes this kind of information is lost beyond recall. Long-term memory is distinguished from the short-term by its enduring character, even for much of a long lifetime. Old people, as we know, can retain and recall childhood memories vividly but somehow fail to enter new memories. This constant, permanent memory can survive when the brain has been cooled or has been in coma or under anaesthetic. Such memory then, must have as a basis some enduring change which, as J. Eccles says, is 'built into the fine structure of the nervous system'. On the basis of what has been said, we can define memory as a property of the nervous system which is effective in the entering, storage, and recall of information.

Attacks on the nature of memory and learning are being made by physiologists, neurologists, and physiological psychologists in many directions. There are those who think the answer lies in a specific reorganization of neuronal associations (the engram) in a vast system of neurones widely spread over the cerebral cortex. Others suggest that the secret of memory and learning lies in the encoding of information in the macromolecules of RNA in the brain. Still others search for the key in the shapes of the nerve cells or in the nature of the synapse. All these approaches, as we shall see, have helped us to understand the detailed activity of the brain at the level of its most basic unit, the neurone. Psychologists, on the other hand, are not concerned with the actual substance of the brain but try to understand and analyse the behaviour of the *intact* animal or human. The psychologist does important spadework for the physiologist and neurologist whose more direct studies may eventually get us further in the understanding of memory and learning. To be fair to the psychologists, however, their specific methods, used on the whole man, may yield important knowledge unobtainable by physiologists on how people of different cultures, or indeed different individuals within our own society, organize their memories, which may differ from individual to individual and from culture to culture.

THE CONTRIBUTION OF THE PSYCHOLOGISTS

Psychologists have told us a good deal about *what* the brain does rather than how it works, and their experiments on iconic or short-term memory are illuminating.

If twelve or sixteen letters of the alphabet are flashed briefly on a screen, we shall be unable to remember all the letters on the screen. If, however, we are told in advance to remember the top, bottom, or middle line then we can do so perfectly. If the line we have to remember is signalled by a simple tone instead of verbally then the signal can be given in advance of the display of letters, simultaneously with it, or after it. If the selective signal (the tone) is given well afterwards then the proportion of letters correctly remembered is about the same as if no selective signal had been given. However, as the time interval between the flashing of the letters and the tone is gradually increased from zero, the proportion of letters correctly remembered does not drop sharply and instantly but dies away slowly. It takes an interval of as much as half a second between flashing and tone to reach a level at which the selective instruction has no effect. Thus, as D. E. Broadbent said, it can be argued that after about a quarter of a second following the visual stimulus there is still some memory trace of *all* the letters flashed on the screen. Some further process can pick off from this rapidly-decaying store any particular part of it and preserve it for further processing in a longer-term memory: but it must do so before the magic interval of half a second is up, since this unselective form of memory for everything that bombards the sense organs is limited by time. In short, the ability to pass on selected information from the short-term memory to a longer-term memory falls off between the critical periods of a quarter and half a second.

We can, perhaps, think of short-term memory as a store holding a fixed number of items which decays from the moment the stimulus arrives, until after about half a second it has gone. If any fresh items arrive they can only be held in short-term memory by knocking out some item already there. An interesting point is that short-term sound memory takes longer to decay than visual memory.

While the process of forgetting in short-term memory can be explained merely by the passage of time, this is not so of the forgetting of long-term memories picked from the short-term store for encoding, and the psychologists have suggested that 'associative interference' might explain forgetting in this case. Briefly this means that old learning is suppressed by new but the new is rapidly forgotten because of the previous existence of the old. In other words, much forgetting is due to the need to remember other things.

If we take a commonsense view of the brain, one of the functions it must perform for us to survive is the prediction of the future, given those events which have just taken place in the present. After lightning we expect thunder; when we see a particular road sign we look out for pedestrians, and so on. In general we learn that event B is likely to to follow event A; but the environment is not static and people are unpredictable, so that event C might follow event A. On this matter, D. E. Broadbent says two things are likely to happen. First, a system which has previously learned to infer B following A is likely to find it very hard to infer C following A. Given a sufficient number of associations of A and C, however, the fresh inference would presumably be established, and we should learn in due course that C rather than B follows A. Second, when the environment returns to its original state and A is followed by B, the confused human being will be inefficient in coping with the situation because he now has incorrectly learned to predict C rather than B.

Clearly we do not think in this rigid, slot-machine way. Any account of brain theory must consider the capacity to learn rapidly and flexibly in a shifting environment where decisions have to be made continuously and statistically.

Broadbent argues that the brain can be thought of as a computer which keeps count of numbers of occurrences of events, so *statistically* it knows the chances of B following A and of C following A. In keeping such a count, any system would need to weight recent happenings (new learning) more heavily than past events (old learning). If this were not so, a change in the environment would not produce an adequate response until the changed system had been in operation for at least as long as the previous state of affairs. A greater weighting for recent events would allow a change to be detected but *not* acted on at once, for A–C rather than A–B might have happened by chance. In other words, the brain is acting like the cybernetic model described in Chapter 14, the basic feature of which is the ability to adjust the rules of its own activity according to the degree of success it attains. It does this not by throwing on-off switches, but by modifying its rules statistically as a result of its own trials and errors.

Such theories and models do seem to fit in with what neurobiologists have discovered about brain function. In Chapter 14 we spoke of a 'wave-front' of activity sweeping in parallel through thousands of neurones in response to the simplest stimulus. Clearly it would be inconvenient if, when a *single* neurone fired, we withdrew an arm or shot out a foot. Brain function, then, is described as 'probabilistic', weighing one course of action against another, and 'parallel' with whole blocks of cells functioning side by side.

Put in another way, the brain has not got to learn or unlearn too quickly or it will associate events which have only come together by chance. It must wait for the events to be repeated so that the probability of learning, say, that C follows A will become steadily greater as the event is repeated, while the probability of B following A gets smaller. As indicated, the brain must be capable of unlearning so that if conditions change it does not remain bound to an inappropriate response. In short, the brain must not learn or unlearn too quickly or it will spin like a weathercock to every changing wind.

Consequent on what has been said, many learning theories envisage something happening in the nerve-webs analogous to the wearing of a footpath by repeated use with a tendency for the grass to grow over the path if it becomes disused. Unfortunately this does not entirely square with the facts of common experience. There are some things that focus our attention at once, others that we rapidly forget even though we try to learn them many times. It seems that those things which rivet our interest and attention and which we feel strongly about *are* recorded by neuronal mechanisms in the brain—probably in the 'interpretive' cortex and the hippocampal system (see Figures 10 and 13) which we shall come to later. On the contrary, the mere passage of nerve impulses which have to do with what we ignore leaves no discoverable 'memory trace'.

From this brief account of the contribution made by some psychologists to the study of learning and memory, we turn now to what the neurologists, physiologists, and biochemists are finding out. One fact which shows the disparity between psychologists and physiologists is relevant here. Physiological studies suggest that the physical consolidation of memories in the brain may go on for half an hour or longer, but the time-scale for the iconic memory is half a second on the psychologist's evidence. Perhaps these physiological changes refer to the events which take place in the psychologist's long-term memory, but as yet there seems to be no sign of a link between psychology and physiology in this field.

THE UNCOMMITTED CORTEX

A direct way of discovering which parts of the brain may be involved in certain definable functions is called ablation—the destruction or removal of a piece of brain tissue to see how an animal's responses might be altered—though not always are *all* possible explanations of the facts considered. Removal or damage at birth of parts of the 'old' brain—the hippocampal system—perhaps prevents memory records from being made, or destroys the mechanism of memory recall.

Another way of mapping the brain is by electrical stimulation of its parts. In response to an electrical stimulus, a conscious patient may have a flashback to the past or the feeling that 'I have been here before'. When the 'smell' area is stimulated a crude odour is experienced.

There are three mechanisms in the brain cortex which are involved in memory and learning, and much has been learned about them by electrical stimulus and ablation techniques. The action of each mechanism depends not only on the cortical area but also on the underlying thalamus and parts of the brain stem. They are thus called 'thalamocortical systems'. Interference with the brain stem causes loss of consciousness, but removal of one or more parts of the cortex deprives a man of only one or more of his functional capacities. Electrical stimulus can map the frontiers of these three 'functional' areas of the cortex. The three mechanisms are:

1. *The motor skills* mechanism developed in both halves (hemispheres) of the cortex. The motor cortex can be traced in the head from the crown downwards and forwards to a point in a little above and in front of the ear (see Figures 10 and 11). When parts of the motor cortex are stimulated electrically with a delicate probe, particular parts of the body move, a toe or ankle and so on. But apart from these elementary, rather crude, responses, no clue is given as to the nature of new skills learned which are really *sensorimotor* skills based on seeing and hearing as well as doing.

2. *The speech mechanism* developed from one half of the 'uncommitted' temporal cortex which is like a blank slate at birth (see Figures 10 and 11). The temporal cortex, as shown in the diagram, is the lobe of the cerebral hemisphere lying under the temple, like the thumb of a boxing glove. On the dominant (left) half of the brain, it is eventually devoted to the learning and elaboration of language. When stimulated electrically the speech mechanism is immediately interfered with, but the interference vanishes as soon as the electrode is removed. If the speech area is damaged in adults, loss of speech occurs, but if this occurs before the age of ten the right half of the brain can take over.

3. *A perception mechanism* in the right temporal lobe, opposite to the speech area in the left lobe, of which the function may be the automatic interpretation of the present by reference to the record of past experience automatically recalled. If an electrode is placed on this right lobe, two sensations might occur in a conscious patient: an awareness of a sudden change in his interpretation of present experience, or a sudden flashback to the past. Both the perception and speech mechanisms are developed from the 'blank slate', both are concerned with specialized learning and memory stores. Understanding the meaning of things (perception), like words and language, has to be learned.

Why are the parts of the cortex described in (2) and (3) called 'uncommitted'? Because of inbuilt nerve connections, vital to survival, the parts of the cortex devoted to hearing, sight, smell, and movement

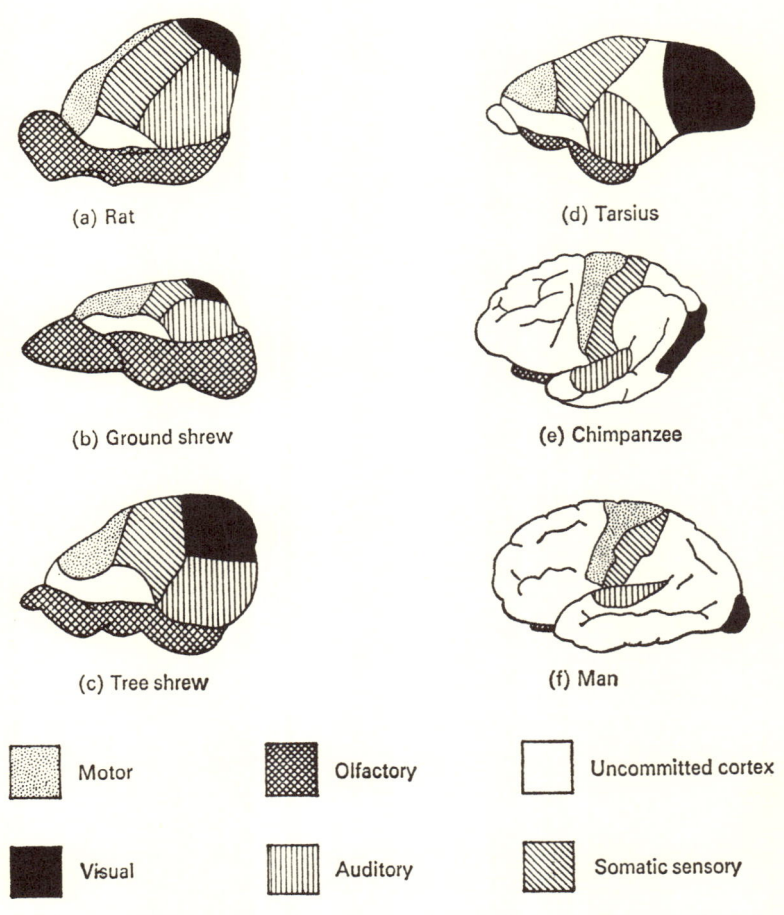

FIGURE 11 Diagram of some mammalian brains, from rat to man, to show the approximate extent of uncommitted cortex as contrasted with sensory and motor cortex. (From Penfield, *Proceedings of the Royal Society of Medicine*, **61**, 1968.)

are committed from birth but the new cortex between the auditory and visual areas (see Figure 11) has to be filled during the first decade of life, as we have said, by learning and experience. W. Penfield called this blank slate the uncommitted or interpretive cortex because it seems to be there to make sense of things seen and heard. Clearly it is important in man compared with other mammals like the rat (whose cortex is mostly motor or sensory), as Figure 11 shows. In man, most of the cortex is neither sensory nor motor in function, and, apart from the speech and auditory areas, most of the back half of the brain is uncommitted or interpretive.

SCANNING THE PAST

When the interpretive cortex is stimulated, as stated earlier, two kinds of response may be produced: a sudden 'signal of interpretation' or a 'flashback'. This latter Penfield calls 'an activation of the record of the stream of past experience'. The first automatically inspects incoming data and judges whether things are familiar, strange, coming near, or becoming dangerous. This mechanism in normal life might scan incoming information to see if it is already familiar by making available from a memory store a limited past. The flashing return of a stream of past experience under the stimulus of an electrode is produced by chance, but in normal life the flashback is *selective* and again used as a scanner. The times that are recalled experimentally are instances when our attention has been focused on an auditory experience, like a voice calling or a piece of music being played, or, not quite so frequently, when we experienced the actions of others. Sometimes sight and sound are mixed, as when children are calling in play.

In summary, the experimental response to an electrode placed on the interpretive cortex is an evoked awareness of thoughts and feelings that once passed through the mind. From moment to moment they parade as fresh and vivid as present experience. In normal life this part of the cortex appears to scan the record of the past to see if incoming data is already familiar or strange or fearful.

But where *is* the memory record? Experimental evidence shows only that there is a scanning mechanism in the temporal cortex, capable of activating the memory thread. According to Penfield the latter is not in the temporal cortex, but according to others it is, and contains a continuous record of our experiences with the emotions felt at the time (J. R. Smythies).

Evidence suggests that part of the old 'smell' brain deep in the centre of the brain, the hippocampal system, has something to do with recording recent memory. Smell and the ability to record and remember smells are important in the animals from which man evolved, and perhaps this memory mechanism, once used in the smell world and once so important to man's ancestors, is used for memory still, in his world of sight, sound, words, and perception.

Penfield, a neurologist, whose hypothesis this is, has removed patients' temporal lobes and hippocampus systems in the course of treatment. When one hippocampus lobe is removed, recording of memory still goes on in the other; but when one is removed and the other is destroyed at birth or by accident the record is not made (or if it is, cannot be recalled). In an engineer whose hippocampus system was removed on one side because of a tumour, and destroyed from birth on the other, and whose speech lobe was partly removed, memory of special skills, reasoning ability, IQ as well as the attention-focusing mechanism on the job in hand, were preserved, presumably in the cortex. He had only to turn away from his drawing board, however, and all memory of what he had done and planned to do was blank. The intact hippocampus system might, then, cause memory to play back like a tape recorder through consciousness but even its *total* loss allows a fair memory of the distant past. Experiments on animals, however, do not lend any support to this hypothesis.

In summary we can say that, whatever the mechanism, only the memory of things that have caught our attention, our thoughts, our emotions, are recorded in the memory system. So selective is the system that only an extremely small fraction of the sensory input into the brain can be recalled in memory. Probably the hippocampus records fairly recent memories and the perception mechanism activates their play-back selectively.

Many questions remain. One large one is: 'What are the basic protoplasmic alterations that make permanent recording of experience and memory recall possible?' A number have been mentioned earlier in this chapter: the engram (see Chapter 11); changes in the junctions (synapses) between neurones; chemical coding and 'reverberating' electrical circuits, especially to explain short-term memory.

MOLECULAR MEMORY

To illustrate the many-sided attack on this problem, three pieces of recent research are interesting: on the theory that 'RNA has memory', on the link between memory and sleep, and on the synapse, which is a front-runner for the principal rôle in the memory process.

The theory that the storage of memory is in the long-chain

molecules of RNA has some experimental support, but at present seems exceedingly speculative. It is well established experimentally that neurones involved in a variety of learning situations do increase their RNA and protein content. Extrapolating from this hard fact, it is argued that perhaps memories are stored as molecular codes in RNA and proteins, just as 'genetic' memory is coded in the DNA of the chromosomes. This may be so, but no one has yet suggested exactly how a chemical memory like this could work in detail, or indeed how nerve impulses can have any influence at all on RNA or protein synthesis, or how such a chemical code can provide the vivid, specific, images and sounds which form parts of memory.

That molecular changes do indeed go on in the brains of animals in a simple learning situation is well established. S. Rose has recently looked at the biochemical changes in the brains of young chicks exposed to a flashing light. His experiments were remarkable in that each chick served as its own control. Their brains were divided so that each half of the brain could be examined in isolation. One eye, and therefore half the brain, was stimulated by the flashing light, and the other eye shaded. Rose found an increase in protein synthesis in the roof of the forebrain of the trained half, suggesting that learning may involve synthesis of protein, perhaps even unique proteins which could be identified chemically.

SLEEP—THE REPAIRER OF SYNAPTIC MEMORY

Sleep seems a far cry from memory, but several pieces of work point to a sleep-memory-learning link. Why do we sleep anyway? One reasonable answer to the layman is to give the brain 'the rest it deserves' after the immense barrage of information fed into it in the day. Physiologists like G. Morruzzi claim, indeed, that those neurones and synapses which *are* the seat of intense activity during the day, and where macromolecular change goes on, need to recover—even though there is a continuing overall activity of neurones in sleep. In short, sleep may be a period of maintenance and repair of the synaptic complexes used in learning and memory.

Perhaps in the young, where new synapses are being formed and the 'wiring' of the brain is changing, sleep is of very great importance. A link, too, between a special kind of sleep, learning, and memory, adds weight to this hypothesis. Figure 12 shows the electrical brain waves

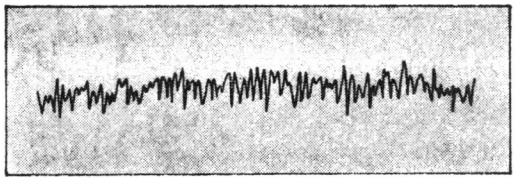

Graph of alpha rhythm as eyes shut.
Still awake, but relaxed for sleep

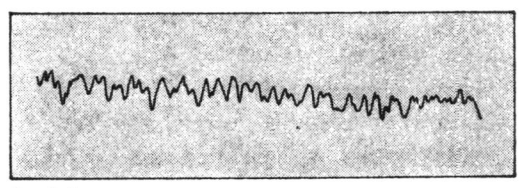

Drowsiness...
registering only slowly and slightly on the graph

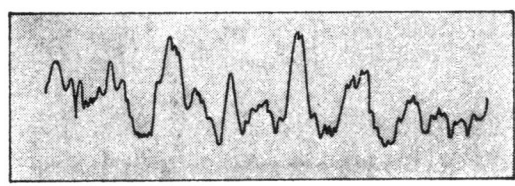

Asleep, shown by craggy alpine skyline
and spindles in the valleys

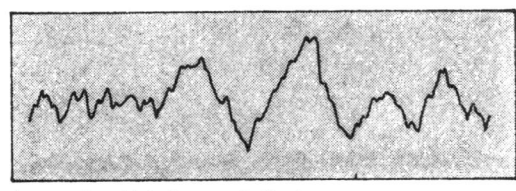

Large, slow 'delta' waves indicate
the final stage of orthodox sleep

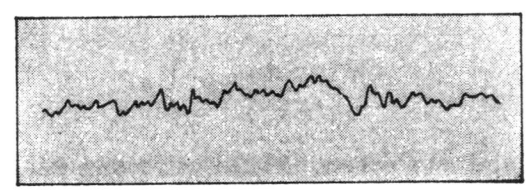

Rapid eye movement sleep
is echoed by vigorous body movement

FIGURE 12 Graphs from *Sleep* by Ian Oswald (Pelican).

associated with different patterns of sleep experienced in a normal night. In REM (Rapid eye movement, see Figure 12) sleep the brain waves are like waking brain waves and dreams occur in this period of sleep. Evans and Newman suggest that dreaming may be likened to the process of clearing and revising computer programmes. Such revision and up-dating of programmes is an essential part of the 'life' of computers and must be done when the machine is 'off-line', that is, uncoupled from the job it is controlling. Likewise the brain requires 'off-line' time for revision and reclassification of information, and this is what sleeping and dreaming are for.

Bits and pieces of information support the idea. Children take more REM sleep at a time when 'input' and memory imprinting must be enormous. By contrast the old, where memory is failing and input small, have less REM sleep. Mentally ill people in whom memory is affected also have less REM sleep.

This memory-sleep theory is further supported by two pieces of recent research. Subjects wearing 'inverting' spectacles take more REM sleep than normal in the period when they are adjusting to the upside-down-world, and patients with brain damage resulting in speech defect learn to speak again more quickly if they have REM type of sleep. But the theory might be too slick. Dreaming sleep may not necessarily be the most important of all. If the deepest kind of sleep ('delta' sleep) is missed as a result of experimental deprivation, on normal nights the loss is compensated for by an increase in this form of sleep as if it had some special restorative property. This links up with Morruzzi's idea mentioned above.

THE SYNAPTIC COMPLEX

We have suggested in Chapter 14 that the wiring of the brain is plastic; that is, neuronal pathways involved in learning behaviour may either grow extra synapses or may redistribute available synapses about the cell body of the neurone or on dendrites. The idea that memory and learning involves modification of the synapse is about eighty years old, and this modern idea of a *physical* change in 'wiring' patterns in response to experience supports the old idea. Perhaps too the synapse is *chemically* modified by learning.

J. A. Deutsch treated rats with drugs which alter the availability of acetylcholine, a chemical 'transmitter' substance which is released from one nerve fibre and carries the nerve impulse across the synapse to another nerve fibre. These drugs which alter the behaviour of the synapse seem to modify memory and behaviour in a predictable manner, perhaps by an increase in the facility with which a particular set of synapses transmit a nerve impulse responsible for learning. Chemical modification of the synapse in learning may well parallel physical changes in wiring patterns.

Whatever the mechanisms of memory and learning they must be compatible with the following facts and ideas:

1. The complex mesh of interconnecting neurones which is widely spread over the cerebral cortex. Any one cortical neurone in this network might not belong exclusively to one 'memory trace', but each neurone and even each synaptic junction could be built into many traces.

2. Memories, in keeping with the complex neurone web, seem to be stored rather diffusely within the brain tissue and are elusive. As we have seen, memory can be damaged by removing parts of the brain, but large areas have to be removed to damage memory severely, and even when this is done long-term memory still survives.

3. Despite the diffuse storage, the mind is one 'organism' and this is shown by the impossibility of speaking effectively while reading, writing, or listening. The mind, then, seems to work as one central co-ordinating register or dictionary rather than from bits of information filed here and there.

4. Human memory and learning thrive on 'associations'. Say a word —'brick' for instance—and at once a collection of memories and associations are summoned, often different for each person. Any theory must explain this commonplace observation.

5. Some neuronal mechanism (the engram, synapse, etc.), whatever it is, makes all that rivets our attention or taxes our thought or emotions *memorable* in a 'memory trace'. And this goes on against a much larger 'background' activity of neurones, nothing to do with the memory. What we pay little attention to apparently leaves little or no lasting record. Something more than the mere passage of a nerve impulse is involved in the establishment of a 'memory trace'. Here once again the selectivity and personal nature of what we want to remember is underlined.

6. Following from the last point is the fact that the neuronal processes underlying learning and forgetting, storage and retrieval of memory traces, are *quantitatively small* with respect to the immense background activity of the cortex, although the highest achievements of mankind, from artistic creation to scientific discovery, are dependent upon them.

7. Long-term memory is durable, even for a life-time, and can survive electric shock, heavy doses of drugs, coma, and cooling until all electrical and most chemical activity has ceased.

If some of the literature to do with brain studies and memory is examined, it can be seen that scientists seem to make little inroads here and there but no central theory of brain function has yet emerged. Some Darwin of our mental life has yet to make sense of the vast and scattered array of observations and experiments on the brain.

17: How Emotion is Investigated

IRRESPECTIVE of race and sex, the components of our brain are stamped out to a basic pattern by heredity, and the function of these basic parts is being slowly and painfully discovered. A point of note is that even though every brain, like any machine, has to have its basic components to work, its uniqueness is striking. This can be judged by the common experience of the infinite and pleasing variety of mankind. Individuality is not easy to pin down, but we know that the genetic character of the fertilized egg is the rock-bottom basis of this individuality and, on top of this and interacting with heredity, is the uniqueness of the environment from the womb onwards. Poor nutrition in pregnancy, as we have seen, can depress a baby's IQ, as can a drab environment starved of talk and stimulating things to do. Brain studies, we know, have revealed that the 'wiring' of the brain changes and develops in the first two years of life, emphasizing the importance of early environment on later psychological development. The subtlety and complexity of the nerve connections might well depend to some extent on the richness, or otherwise, of the environment.

Many of the brain components like the visual, auditory, and motor cortex, as we have seen, are committed for instant action at birth, but others, such as the speech and perception areas, are like blank slates ready for the inscriptions of specialized learning and memory. This fact emphasizes the importance of early experience, perhaps up to ten years old in this case.

Further understanding of the brain as a machine has come from the investigation of its more primitive parts, the 'old' brain—old in an evolutionary sense—which is deep in the middle of the skull, overlain by the newer part, the cortex. The techniques, which are relatively crude and undiscriminating, involve plunging fine wires through the cortex into the central structures (the cortex is very little damaged in the process); by means of these implanted electrodes, electrical stimulation of small areas of the brain has been possible and correlated with the behaviour of the animal. Later the animal is killed and the exact position of the tip determined by dissection. Equally, a current could be sent through the wire to destroy cells at the tip and the effect on behaviour observed. Some of the most striking experiments on these lines have been done by Delgado, who uses a special radio transmitter/receiver which keeps him in direct contact with electrodes implanted in special areas of the brain. By throwing the switch of the transmitter, the receiver lodged in the animal's skull activates the wires to a particular part of the brain and a particular behaviour is elicited. One dramatic use of this electrical control of behaviour was carried out with a fierce bull whose brain has been wired with electrodes both in the aggressive centres of the hypothalamus (see Figures 10 and 13) and in the motor centres in the cortex. Delgado entered the bull ring armed with a transmitter. At first the bull behaved normally and began to charge, but when it was a few feet away from Delgado he threw the switch of his transmitter. The bull stopped abruptly and meekly turned away. This technique has been used mostly with rats, monkeys, and cats, but already it is being used with people whose behaviour is seriously disturbed by epilepsy.

The inner kernel of the brain we are considering here is called the 'limbic system', what we have called previously the primitive, or old, brain. Basically we know that it is involved in emotional behaviour, and besides the hypothalamus it consists of a number of intricately connected structures which include some already mentioned in previous chapters —the hippocampus, amygdala, reticular formation, and various regions of the cerebral cortex like the temporal lobe.

Investigation of the functions of the various parts of the limbic system involves techniques that are difficult to use. The results they give are hard to interpret with certainty, and since much of the research is done with rats, cats, and monkeys, extreme caution must be used in transferring such interpretations to man. Some of the research and its difficulties is considered below.

1. Let us consider the hub of the limbic system, the hypothalamus, which in man is as big as a damson, and in rats, of course, much smaller. Nevertheless, except for size, it has quite a similar structure in all mammals and the hypothalamus does not appear to differ much in man, cat, and rat. Its relatively tiny mass is very complex and includes at least twenty 'centres', some controlling temperature, others aggressive and feeding behaviour, while yet others inhibit

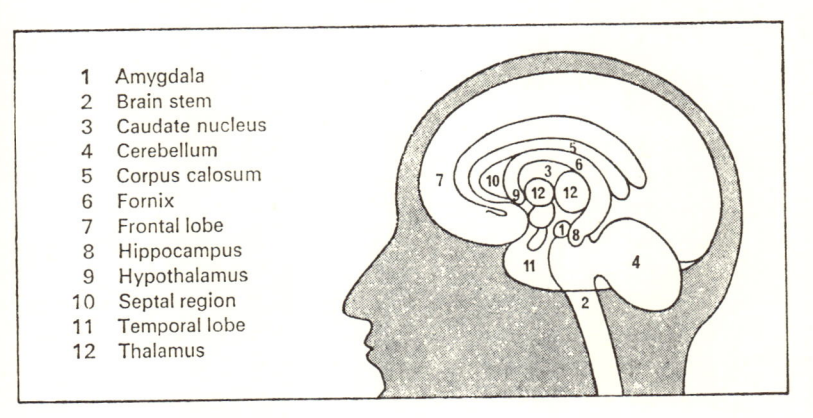

1	Amygdala
2	Brain stem
3	Caudate nucleus
4	Cerebellum
5	Corpus calosum
6	Fornix
7	Frontal lobe
8	Hippocampus
9	Hypothalamus
10	Septal region
11	Temporal lobe
12	Thalamus

FIGURE 13 A map of the brain.

an animal's feeding when it is satiated. A few years ago the public were fascinated by the discovery of a 'pleasure' centre, but balancing this is an 'unpleasure' centre. Really these centres are better seen as a very basic approach-avoidance mechanism which takes its cues from various higher systems, and is capable of evoking appropriate bodily responses. The complexity of these centres is shown by Miller's experiments with rats. If a small area is destroyed on both sides of the hypothalamus, the animal will stop eating and starve to death. Conversely, electrical stimulation in this same lateral area will cause an animal which has just eaten to satiation to eat voraciously. Such stimulation also makes the animal go through food-seeking behaviour. It is as if the rat is pushed by electricity into certain patterns of behaviour. Clearly the results of destruction and of stimulation agree in showing that the lateral hypothalamus is significantly involved in food-seeking behaviour. As a further illustration of how complex this tiny area is, damage to the lateral hypothalamus causes animals to stop drinking water as well as to stop eating food. And electrical stimulation of certain parts of the lateral hypothalamus will cause rats that have drunk to satiation to start drinking again. Stimulation of other areas of a cat's hypothalamus by small electrical currents can send it off into an attack of ungovernable fury, accompanied by attack movements of the body. When the current is switched off this 'sham' rage collapses.

2. Not only can electricity trigger certain forms of behaviour in rats, but so too can minute drops of substances normally found in the brain, or synthetic substances resembling these. Alan Fisher showed that specific chemicals dripped on the rat's hypothalamus affect specific groups of neurones in specific ways. This probably means that the brain has a complex chemical pattern brought about by electricity triggering a flood of chemicals.

Fisher dripped a *male* hormone through a hollow needle into a particular part of the hypothalamus and found that it caused female behaviour—nest building, carrying pups to the nest—in both male and female rats. When the same hormone was applied to a slightly different area it caused male rats to mount non-receptive females and to go through all the normal sexual pattern, including ejaculation. Female rats behaved similarly with the exception, of course, of insertion and ejaculation. When the tube was shifted to a point between the two positions, mixed behaviour resulted: a male might mate with a female and then carry strips of paper to build a nest. All these effects happened within twenty seconds to five minutes of injecting the hormone and were quite specific, not being produced by injections in other parts of the brain. Curiously, the female hormone, oestradiol, which is chemically like the male hormone testosterone, produced similar but less potent effects.

Fisher's work thus showed that the programmes for male and female behaviour are present in the brains of both male and female rats. These patterns can be triggered by the presence of male and female sex hormones acting on the hypothalamus. This is a subject we shall return to in Chapter 18.

One interesting technical point may be made here. An electrode plunged into an area of the brain affects the synapses of the cells in that area by electrical stimulation, but it also affects nerves passing through that area so that the behavioural results are difficult to interpret. Electrical stimulation, then, affects the telephone cables as well as the switchboard, to use a loose analogy. Chemical stimulation, on the other hand, might produce results on behaviour which are more specific to given areas and may act on synapses without affecting fibres of passage.

3. Another area of the limbic system, the amygdala, when stimulated in the cat's brain by electricity, causes a 'searching' response. The cat stops what it is doing and raises its head, its eyes open wide, its pupils dilate, and its ears prick. The whole attitude is expectant. More intense stimuli to one part of the amygdala can give rise to an aggressive/defence reaction, in another to a cowering response. Perhaps like the hypothalamus, the amygdala contains a number of centres, carefully balanced and linked with the hypothalamus and reticular formation to do with rage-aggression-defensive responses. In Rhesus monkeys destruction of the amygdala tames the animals. In man stimulation of the amygdala can lead to a lowered rage threshold and outbursts of aggression, while destruction for therapeutic purposes of parts of the amygdala can improve the behaviour of excitable, aggressive, and destructive youths.

4. From what we have just said about the hypothalamus and the amygdala, it is clear that they are 'Chinese boxes' of complexity. Indeed, the limbic system seems to be a network of areas, spread out yet interlaced and minute. This diffusion of function may well be adaptive, since a tiny amount of damage from infection or a blood clot would not completely stop a vital function. But clearly it makes the investigation of the organization and function of the inner brain and the consequent interpretation of behaviour very difficult. Indeed a good many conflicting results on behaviour have been obtained by the techniques mentioned.

5. A further complexity is that the limbic system is characterized by a number of 'closed loops' which provide the structural basis for a positive 'feed-back' system as well as allowing for alternative pathways for nerve impulses to be sent through the various

components of the system. A feed-back system (see Chapter 18) works like an ordinary thermostat and tends to a balance point. If one structure in the limbic system does not maintain a proper level of secretion, another may respond by raising or lowering its activity. Complication arises once again in the interpretation of experimental results with electrodes or added chemicals. Injection of a chemical (carbachol) which causes drinking into *any* of the limbic structures of water-satiated rats will cause them to drink. Where is then, the specific 'thirst' centre?

6. How does all this research into the centre-brain of animals apply to man? Twenty-five years ago the hypothalamus was known as a small, obscure region. But now, mainly on the basis of animal experimentation but with some chemical supporting evidence, it is thought that there are centres within it like those of the rat or cat for controlling various types of behaviour. Two of these are the 'feeding' centre, lying laterally, which we have referred to previously in rats, and the 'satiety' centre, lying 'medially.' These two centres act in a 'pull-push' way; nerve activity within the feeding centre leads to feeding behaviour, and nerve activity in the satiety centre shuts down feeding activity. Destruction of the feeding centre in rats causes death from starvation, destruction of the satiety centre leads to massive obesity. (A tumour of the pituitary gland in man probably gives rise to the rare disease Fröhlich's syndrome, leading to the sleepy, fat, boy famous in *Pickwick Papers*. The tumour affects the hypothalamus which lies above the pituitary.) Acting upon this 'appestat' in the hypothalamus are various mechanisms like blood sugar changes, so that the brain mechanism can respond to cues from within the body. However, in man the natural functioning of the appestat can be undermined by a host of human qualities: unhappiness, anxiety (perhaps the result of a disturbed home or a row with the boss) can make us grab for sweets or cakes as a sop. Or the false baits of lovely foods in glossy magazines can set our mouths watering when we are not hungry. This idea brings us to an important point about the human limbic system. In animals this is a prime mover in emotional behaviour. In man, electrical stimulation, surgery, and various abnormalities of the hypothalamus, show that it can reorganize, recode, and redistribute relatively unorganized emotional behaviour patterns, together with its limbic connections, but it must take its cue from

the thalamus or cortex. In man these structures really know about the dangerous, the embarrassing, and the beautiful. The hypothalamus, in balance with the limbic system, communicates these higher decisions and influences to the stomach, blood vessels, sweat glands, and so on, as we know well enough by common experience, through the autonomic nervous system. In man, then, the hypothalamus with its limbic connections is only an accessory and not an instigator of emotional behaviour. As Elliott says 'it has neither the data nor the neurone array to recognize experiences with emotional qualities. It is far less in touch with the world . . . than is a free-living worm (and a worm's emotions can hardly be very intense). To equate it with emotion itself is absurd.'

From what has been said about the techniques of studying the old brain it seems clear that we now have the tools to tease apart much of the anatomical, chemical, and physical basis of *crude* emotional behaviour in man, but subtle rational (and irrational) distinctly human behaviour is a function of the whole brain with its genetic and neuronal individuality. We are a long way from understanding or controlling human behaviour.

THE SHOT-GUN EFFECT OF DRUGS

Perhaps the word 'balance' sums up the activity of the limbic system which by feed-back mechanisms controls certain types of behaviour through the autonomic neurones and glandular system which we shall deal with presently. This naturally leads on to the use and development of drugs for the anxiety and emotional disturbances which are now so common and are associated with malfunctioning of the autonomic system.

While doctors and scientists have worked wonders to develop drugs that act, say, on the amygdala (possibly the anatomical source of anxiety and aggression) the action of such drugs is hit or miss, more like a shot-gun than a target rifle. From what we already know about the limbic system—its overlapping actions, its link with higher centres, the fact that in the rat the limbic structures are separated by less than a millimetre, and the relatively undiscriminating methods of brain exploration at our disposal—how can we expect drugs to produce bull's eye effects? Yet it is by these methods and others still to be developed that a more rational foundation for the discovery of new drugs and the use of present ones, for treating certain forms of mental disorder, will be developed.

The crude knowledge we have now is a warning and a hope about drugs. As we have said, the word 'balance-point' sums up the activity of the centres of emotion and, like the action of a thermostat, the balance-point can be set higher or lower. Hard drugs might well set the 'thermostat' permanently towards depression, so that the taker of heroin needs his drug to restore a normal balance-point. It is significant, too, that withdrawn schizophrenics, after stimulation of their 'pleasure centre' in the hypothalamus, become normal and outgoing for a while.

DRUGS AND SYNAPSE

One field of discovery which might hold the key to the solution of many human ills concerns the synapse—the sub-microscopic junction between one neurone and another.

Synaptic thresholds have been mentioned previously, and it is known that the electrical impulse passing down a nerve fibre releases a flood of chemicals called transmitter substances which bridge the gap to the next neurone, either stimulating it to fire and transmit an impulse, or inhibiting it. Perhaps nervous fatigue means, in chemical terms, an exhaustion of transmitter substances and an accumulation of waste products at the synapse, resulting in high thresholds. A strong cup of coffee containing the drug caffeine lowers the thresholds so that the nervous system, though short of transmitter substances, can help us to burn the midnight oil a bit longer. Smoking a cigarette which allows nicotine to enter the body raises the synaptic threshold (to slow reactions) and might help to calm frayed nerves. Adrenalin lowers the thresholds of many synapses (and thus speeds up reactions) in emergencies like being chased by a fierce bull. Perhaps clinical depression raises the threshold to produce slow thinking and talking (as does alcohol) and drugs to control depression seem to lower the threshold of whole blocks of cells. Clearly, then, understanding the control of synapse chemistry could be a blessing to mankind, but like any other scientific discovery it could be put to evil use.

FITTING THE JOB TO THE MAN

Another field of discovery is not chemical or physical at all, but to do with the human use of human beings: making proper use of human variety and talent, and fitting the job to the man and not the man to the job.

The early chapters of this book emphasized the genetic uniqueness of the individual, and brain studies confirm this fingerprint individuality. Schools recognize more and more this uniqueness, and their methods of teaching change to 'light fires rather than fill pots'. Many are moving over to investigational methods in teaching where pupils learn to assess data, make hypotheses even though penny-sized ones, and go on to test hypotheses by new experiment and observation. These methods help to emphasize and make more use of the unique qualities of the human mind by allowing decision and choice. This unique quality is the more important now that the machine is taking over the slavish jobs and time is left over to use brains at all levels, for example in decision-making processes. Decision-making used to be left to the bosses; now, at every level and in every field—in the road maintenance system, in the Post Office, on the factory floor—people are having to make more decisions and use more complex machinery responsibly. Unfortunately, in a society where machines are replacing men, this has often meant the disintegration of old, satisfying, craft processes into repetitive jobs with the mass use of human beings. We are told that at least 35 per cent of the 400 million working days lost annually through sickness is of psychological ('functional' as the doctors call it) rather than of organic origin, and perhaps this is because of a rebellion of the mind against jobs which depersonalize a man.

Today most young people come from better homes, better schools, and by and large are better fed, clothed, and physically more fit, than any previous generation. Many from this background enter dirty, noisy factories where the monotonous routine of repetitive work is more suited for the mentally retarded than the fit. They look forward to independence but find drudgery.

Job satisfaction and the feeling of being wanted and useful is powerful medicine for poor performance, high labour turnover, and perhaps much sick absence. If job satisfaction was the norm, many of the drugs for anxiety and psychosomatic illnesses would be unnecessary.

MAN—A COMPLICATED DETERMINATE MECHANISM

This section on the brain emphasizes two points: first, its machine-like nature; second, its unique quality in that each brain is different. Machine-like in that the raw materials for thinking and doing are sugar, oxygen, and electricity, and the basic components of every brain are made in the same genetic mould, yet a machine so complex that we are only on the fringes of understanding it. Unique, because each brain not only differs from other brains genetically but also in its 'wiring', its memory stores, what *it* regards as important for memory purposes, what opportunities it has been given in childhood—and before childhood in matters of nutrition—and the level of the synaptic thresholds.

The notion that we are after all only machines, albeit unique and complicated ones, leaves us weak at the knees. But this is the only working hypothesis a scientist can make: that man is a very complicated determinate mechanism. Such a hypothesis in the long run can be *disproved*, whereas if free-will is invoked in human behaviour progress in understanding man will be blocked. For free-will implies that there is an event in the brain (free choice) which is in principle unobservable and yet which determines behaviour. No amount of science can investigate such a causal event in other people. Nevertheless, what was said on page 5 is worth recalling: knowledge of the controlling systems of our behaviour helps us to understand ourselves and might perhaps help to reduce misery and increase happiness by medical advance; but none of us could live our lives sanely and sensibly and with humour if we did not adopt other hypotheses—in the form of free-will and, by a majority, a deity. But the discovery of controlling mechanisms by the scientist and the concept of free-will are separate, the latter carrying us well beyond the scientist's field in which the methods of the scientist can be productive.

Discoveries such as those described represent the fruits of patient observations and experiments on man and other animals over the past fifty years. Clearly, on present evidence, human behaviour does not seem to be as 'free' as was once thought, but tethered by interaction of unique inherited and environmental influences. This statement is backed up by the effects of hormones on brain development and behaviour, a field considered in the next part.

Part 3: Chemistry and Temperament

The chemical signals discharged into the blood from the endocrine glands can affect the basic temperament laid down by heredity, but it is difficult to tease out and isolate their effects on individuality from those of the nervous system and brain. The leader of the endocrine orchestra is the pituitary gland, as big as a hazel nut. Its cues are taken from the hypothalamus, a part of the brain, which acts through the autonomic nervous system and the pituitary gland. Brain, nervous, and endocrine systems are closely interlocked to maintain health. Disease rocks the three-legged stool. A marked overactivity of the adrenal cortex may cause women to develop masculine characteristics because it produces steroids with the properties of male hormones. When the anterior pituitary gland functions poorly, growth of the body is impaired, puberty delayed, and development of the reproductive organs may even fail altogether. All these physical diseases of chemical imbalance are found to affect personality. Indeed speculation from some historical evidence suggests that pituitary failure in Napoleon at the age of forty marked the turning of his tide of fortune as a military leader, though other illnesses dogged him. The change in the personality of Henry VIII has been attributed to thyroid failure by one authority and to gout, an inherited metabolic disease, by another.

On firmer evidence, the waxing and waning tides of steroid hormones during the normal menstrual cycle can affect the mind of some women strongly. Crime, suicide, and accident rates rise at the premenstrual stage.

Perhaps the most powerful imprint that hormones (and other biochemical substances) leave on the mind takes place before and soon after birth. At these stages they can affect intelligence, lay down the pattern of future sexual behaviour which develops at puberty under the influence of sex hormones, and imprint the sex difference on the hypothalamus, a difference which triggers the cyclic pattern of sex hormone release in the girl and a non-cyclic pattern in boys.

In this book we have seen that chromosomes and genes do affect personality. Indeed they are bound to, for all our mental and physical properties are to some extent limited and prescribed by the chromosome outfit of the fertilized egg from which we develop. We have seen how the genes might act through the brain and the nervous and hormone system to influence personality, but how they interact to produce the complexity, variety, and subtlety of human nature is something that will take us much longer to learn.

18: The Endocrine Glands: an Introductory Sketch

IF we think of the cerebral cortex as the inner cabinet of the brain deciding on policy, then the hypothalamus, the brain segment we have talked about in previous chapters, might be the Civil Service with its network of centres or 'Ministries' which deal with the day-to-day running of the strategies laid down by its masters in the cortex. These centres, as we have seen, control primitive reactions to fear, anger, hunger, and sex, and work through two closely interlocked systems which influence body and behaviour. These are the autonomic nervous system described earlier in Chapter 11, and the endocrine or chemical hormone system.

Hormones are chemical signals sent in the blood, and together with the nervous system they co-ordinate bodily reactions: nerves tend to give quick results, while hormones act in a more leisurely way but with longer-lasting effects. Both sexes have the same hormones but put some of them to different uses.

The hormones are discharged from endocrine glands into the blood stream, directly from the cells where they are made. In this they differ from the secretions of the exocrine glands which have ducts. The salivary glands are examples of exocrine glands, as are the mammary glands in the breast.

In response to information fed back to it in the blood stream, the amount of hormone an endocrine gland secretes can be adjusted, just as a constant temperature is maintained by a thermostat. This principle of 'feed-back' is an important one to which we shall return later in this chapter.

THE HORMONE-GENE MAP

Where are the endocrine glands and what do they do? Figure 14 shows their position in the body. Most of them are small and all of them could be held in the palm of one hand.

It is not the purpose of this book to describe the endocrines gland by gland, for our main purpose is to see what effect some of these chemical signals have on our behaviour and personality. At this point, however, it might be useful to state a few facts about the endocrine glands, together with any evidence of their genetic control.

1. The pituitary gland, about as big as a hazel nut, lies at the base of the brain. The anterior, or front lobe, issues its chemical instructions to other 'target' endocrine glands: the thyroid, the ovary, the testis, and the adrenal cortex. Besides this it secretes a growth hormone (GH) which acts on all the cells of the body. Disorders of the cells producing GH in childhood can give giants (the 'Alton' giant was about nine feet (three metres) tall) or dwarfs. (see Figure 15). In adults when bone growth has finished, the soft tissues and face bones are distorted strangely by GH disorders leading to acromegaly. Besides GH the anterior pituitary releases the adrenocorticotrophic hormone (ACTH) which acts on the outer part (cortex) of the adrenal glands and causes it to release its 'steroid' hormones (see (4)) in a balanced way by a servo-mechanism which is described on page 165. The two gonadotrophic hormones released by the anterior pituitary, follicle-stimulating hormone (FSH) and luteinizing hormone (LH) act differently in men and women. The former produces follicles (see Figure 17) in the ovaries of women and spermatozoa in men. LH (called interstitial cell stimulating hormone, ICSH, in men) triggers ovulation and makes the ovary produce progesterone in women, while in men

The Endocrine Glands: an Introductory Sketch 155

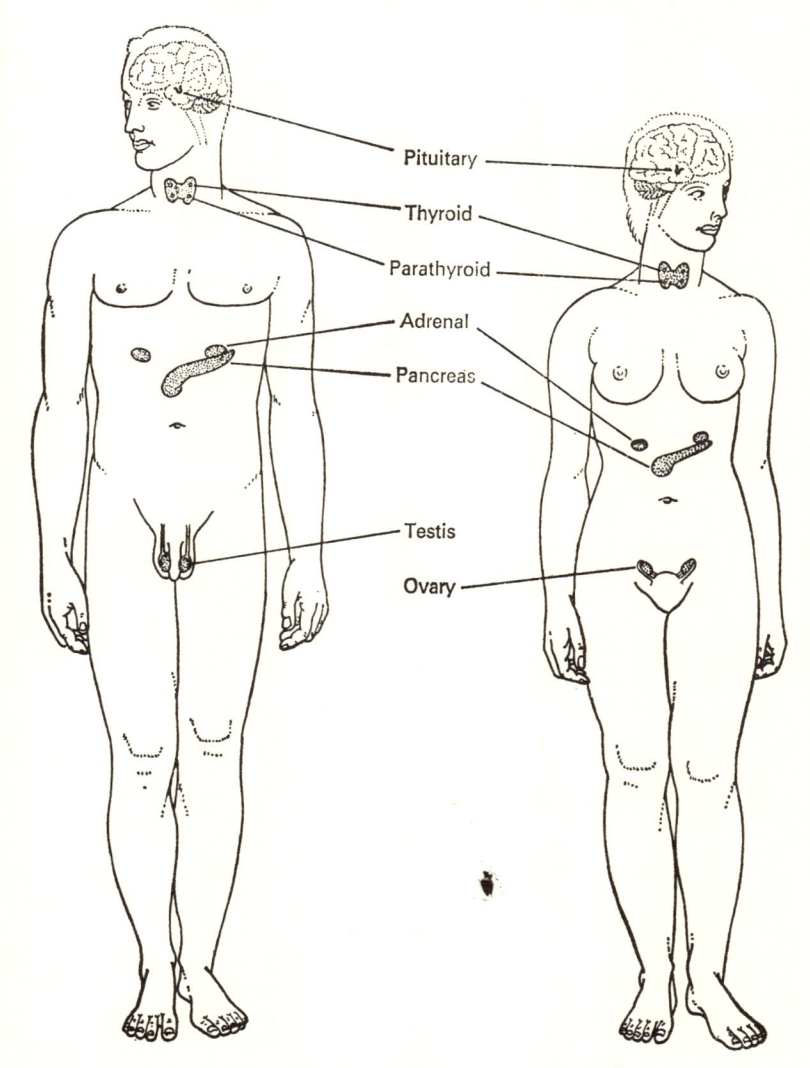

FIGURE 14 The position of the endocrine glands in the body.

it controls the secretion of testosterone. This underlines the point made earlier: that *identical* hormones are put to different uses in male and female.

Thyroid-stimulating hormone (TSH) is another of the battery of chemical signals released by the anterior pituitary, and controls the

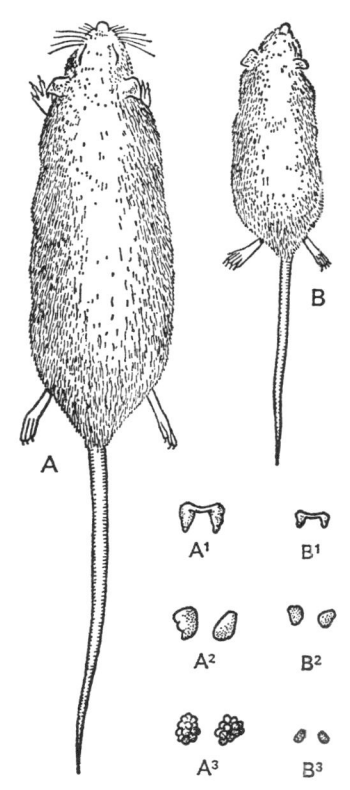

FIGURE 15 Effect of pituitary hormone on growth may be demonstrated in the rat. The pituitary was removed from one of two littermates thirty-six days after birth, at which time both rats had the same weight. After several months the normal animal (*left*) had tripled its weight and had matured while the other (*right*) had gained little weight and was maturing much more slowly. At left are the thyroids (A1), adrenals (A2), and ovaries (A3) of the normal rat, and at right are glands (B1, B2, B3) of operated rat. (From Zuckermann 1957.)

activity of the thyroid gland by a beautifully balanced feed-back mechanism described on page 165.

Interestingly, the hypothalamus imposes its rule on the anterior pituitary by secreting 'releasing factors' from its nerve endings into the blood. These hormones are in turn carried to the anterior pituitary and control *its* output of LH, FSH, ACTH, and GH. By a feed back mechanism described on page 166, these pituitary hormones regulate the secretion of the releasing factors. There are indeed wheels within wheels in the endocrine system.

The posterior, or back, lobe of the pituitary stores and releases two hormones made in the hypothalamus. One of these affects blood pressure because it can cause constriction of the walls of the small blood vessels. In addition it controls the volume of water lost through the kidneys. The other causes milk expulsion in response to suckling and also stimulates contractions of the uterus in labour. The median, or middle, lobe makes a hormone which can darken the skin in amphibians because it expands the pigment-bearing cells of their skin. Its function in man is uncertain.

The pituitary gland has been called the leader of the endocrine orchestra. It is clear, however, that the hypothalamus is the real conductor, with the pituitary playing first fiddle and the rest playing second fiddles.

2. The thyroid gland, below and at the sides of the larynx, secretes thyroxine and triiodothyronine, but how the effects of these hormones are parcelled out in ordinary life is not clear. Their basic function is to control combustion in the body, regulating the chemical 'furnace' and thus mental and physical activity—normal, sluggish, or excessive.

Quite independent of this regulating function is the importance of thyroid hormones on growth and development. Removal of this gland will inhibit growth in a young animal, delay long-bone formation, and prevent proper development of the reproductive system. Its effects on the development of the brain before and after birth will be mentioned on page 170, but are profound and far-reaching on personality. The wide-ranging action of thyroid hormones includes increasing the blood supply to the breasts, thus affecting milk production; control of water and salt balance in some tissues; and the making and destruction of protein as well as modification of the fat balance in the blood.

Numerous genetic errors have been discovered in the manufacture of thyroid hormones. At least five defects are known, each of which can by itself produce goitre, with consequent effects on personality. Possibly a double recessive gene can lead to thyroid disease, but people with only one such recessive gene (heterozygotes) who have a lowish thyroxine output are not uncommon. They seem able to maintain normal thyroid function under usual conditions but

develop goitres under stress. Such genetic defects acting through hormones might not affect behaviour or personality except under stress.
3. The parathyroid glands, four beads of tissue embedded in the back of the thyroid, influence growth and calcium and phosphate metabolism, vital to life. Nearly all the body's calcium is contained in bone, which is not a static substance and is continually being built up and dismantled. From the calcium reservoir in the bone, calcium is withdrawn to be used in muscle contraction, nerve transmission, blood clotting, and, of course, in pregnancy, for the growing baby in the womb.
4. The adrenal glands sit like cocked hats on top of the kidneys. They are vital to life. Their outer rind or cortex produces half-a-dozen different 'steroid' hormones (steroid because of a particular kind of molecular shape). Besides regulating carbohydrate metabolism and sodium balance (life is impossible without this dull-sounding substance) the steroids adapt the body to prolonged stress.

A number of inherited defects which affect the complex, step-by-step building-up of the steroids, are known in man. A double dose of a recessive gene may lead to excessive secretions of androgens by the foetal adrenal cortex, and thus to 'virilization' of female and male children even at birth, so that girls resemble boys. The personalities of untreated children, especially girls, must suffer considerably.

Other genetic defects, again possibly double recessives, lead to a low output of the steroids which adapt the body to stress. What happens to the unfortunate people who have an unusually limited capacity, say due to one 'bad' gene, to synthesize an important steroid when they encounter a highly threatening personal situation? Are these heterozygotes in a position to stand up to the daily business of stress? If so, does the long-term shortage of hormone have a bad effect on brain function so that the facing of a difficult personal problem becomes more difficult? The gene-hormone-personality axis must strongly affect some unfortunate people here.

The adrenal medullae, as we know, secrete adrenaline and nor-adrenaline which reinforces the sympathetic system in meeting sudden emergencies. They have been called the 'glands of emergency.'

Adrenaline/nor-adrenaline balance might well affect our personalities towards anxiety on the one hand, and rage and anger on the other. Some fascinating work described by D. H. Funkenstein shows that aggressive animals like the lion have high amounts of nor-adrenaline but that adrenaline predominates in the rabbit, which depends for its survival on flight. Domestic animals and wild animals that live social lives (baboons) have a high ratio of adrenaline to nor-adrenaline. Man has within him the lion and the rabbit. He is born with the capacity to react with a variety of emotions, but inheritance, interacting with early childhood experiences, may largely determine whether he will be a rabbit or a lion under stress.

5. The 'islets' of the pancreas produce insulin, which we know speeds up the rate at which the tissues use sugar. Diabetes is a well-known inherited disease, perhaps due to many genes affecting the 'islets', or to a dominant gene which produces an 'antagonist' to insulin.

6. The gonads—ovaries and testes—besides producing eggs and sperms, regulate sexual development and cycles, disorders of which can have profound effects on behaviour. Oestrogens and progesterone are released from the ovary, and androgens, of which testosterone is the most important, from the testes. These we shall deal with in the next chapter, but testosterone is the major cause of the increase in size and strength of the male muscles at adolescence. 'Oestrogen' is a loose term for several hormones, the most important being oestradiol responsible for the development of a girl at puberty into a woman. It is interesting that the glands of each sex secrete both male and female hormones and it is primarily an excess of male hormones over female hormones, or vice versa, which controls differentiation. We have already seen that absence of an X chromosome (XO) causes sexual abnormality and that maleness is imposed by a Y.

LINKS BETWEEN MIND AND MATTER

We know that the hypothalamus is the head of the sympathetic branch of the autonomic nervous system which prepares the body for attack or defence, with the help of the secretions of the adrenal medullae. If we are frightened by a response arising in our higher centres—the cortex—an alarm is sent to these glands by the hypothalamus, through sympathetic nerves. In response to the hormones the heart races to rush blood to our muscles, our blood pressure rises, our blood sugar increases to supply fuel to the muscles, and we breathe hard to fill the blood with oxygen for action.

The hypothalamus also controls the other branch of the autonomic system, the parasympathetic whose action is to conserve the body and build up its inner resources. Its actions bring peace to the body. The heart slows, breathing is peaceful and slow, sleep is deep, blood pressure falls. In ordinary circumstances the action of the sympathetic and parasympathetic balance.

All this establishes a mechanism for a commonplace of human experience: that thoughts, sights, smells, words, sounds, can produce bodily and mental changes. A good deal of our recognition of emotion in ourselves is by an awareness of these bodily changes. Think about food and your mouth waters. Words or thoughts can prepare sex organs for function or cause a blush or shiver. Strong emotion can play havoc with the stomach and heart. The mechanism for these mind-matter interactions is: cerebral cortex → hypothalamus → autonomic nervous system → target endocrine glands (adrenal medullae). But we need to add to this a second pathway from the hypothalamus: the pituitary gland. Its effects on the body and mind through the 'target' endocrine glands of thyroid and gonad are profound, as we shall see.

The hypothalamus and pituitary, then, one walnut-sized, the other pea-sized, form the centre for much of the life and loves of man. They communicate one way to the cortex, where the real decisions are taken, and the other way to the body through the closely-meshed endocrine and autonomic nervous systems. R. Greene summarizes this in the diagram on page 161.

What is high-lighted here is the close interlocking of autonomic and endocrine systems to maintain the balanced emotions and actions of the healthy body. These balances are ready at a moment's notice to meet the needs of life by either diminishing or increasing activity ('feed-back' systems are important here and will be considered soon).

The close association of endocrines and autonomic nervous system has been noted at the central level: the link between the hypothalamus, a part of the nervous system, and the pituitary gland. We have noted, too, the fact that the 'fight and flight' hormones, adrenalin and noradrenalin, are triggered by impulses of the sympathetic system, and these hormones in turn react upon the nervous system, heightening

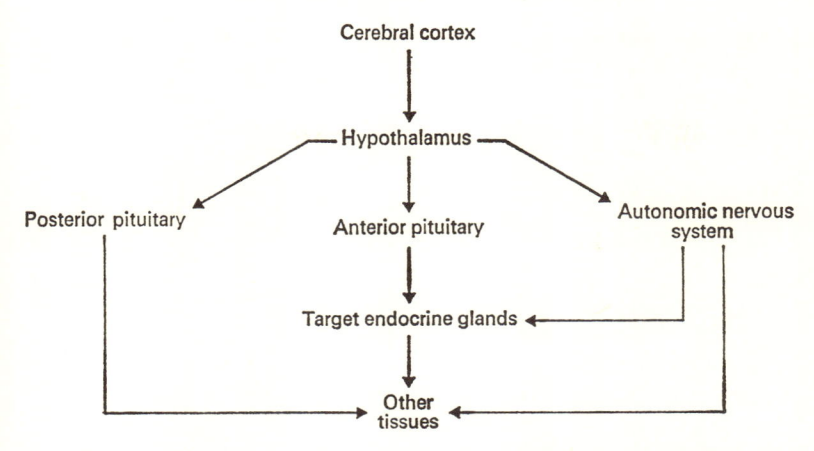

The Endocrine Glands: an Introductory Sketch

```
                    Cerebral cortex
                           |
                           ↓
              ┌───── Hypothalamus ─────┐
              ↙            ↓            ↘
Posterior pituitary  Anterior pituitary  Autonomic nervous
    │                      │                 system
    │                      ↓                   │
    │              Target endocrine glands ←───┤
    │                      │                   │
    │                      ↓                   │
    └────────────────→  Other  ←───────────────┘
                       tissues
```

From Greene, R. *Human Hormones* (1971) World University Library.

its response to stimulation. In addition, the 'setting' of the level of responsiveness of the autonomic nervous system is related to quantity of the thyroid hormone, thyroxine, circulating in the blood. Probably this level is genetically determined and disease can notch the setting up or down, with profound effects on body and mind. Most interesting in this link between nerves and glands is the fact that not only is nor-adrenaline released by the adrenal medullae, but is also important as the chemical transmitter between many nerve endings, triggering chains of nerves to 'fire', and between the endings of the sympathetic nerves and the organs they 'influence'—the muscle walls of the blood vessels, for instance.

HORMONE BALANCE

A number of points may puzzle the reader. Why is it that nature, usually so economical, should arrange for an endocrine gland to pour its secretions haphazardly into the blood to affect one particular target organ, say the ovary? Perhaps it is because every hormone has a job to do in every tissue of the body. Hormone action may be the same for all tissues in general terms, but modified in detail because the cells will have different enzyme systems, different proportions and types of protein, fats, and so on. Insulin, the major control for sugar metabolism, secreted from the pancreas, and thyroxine, produced by the thyroid and acting as a kind of governor, determining the speed of physical and mental activity, affect every cell. Perhaps other hormones in some, as yet obscure, way affect all cells.

Next we can ask, does the importance of hormones lie in each having its own particular job to do or is the balance of the whole hormone and bodily system of greater significance? All biological knowledge of man points to the fact that the welfare of body and mind is dependent on systems of fine chemical checks and balances, in this case different members of the endocrine circle working in harmony with the organs of the body. Should this harmony be disturbed, the health of body and mind soon suffers.

SUGAR BALANCE

Consider the metabolism of sugar. For good health the supply or level of sugar in the blood must be held within close limits. Four factors enter here:

1. the rate of use of sugar for the supply of energy to the tissues;
2. the rate of absorption of sugar by the blood from the intestines and of re-absorption from the fluid filtered by the kidneys;
3. the release of sugar from the carbohydrate-storing tissues (for example, the liver and muscle);
4. the formation of sugar from fats and proteins.

In all these processes hormones play a part.

Greene, writing about the machinery which regulates the level of blood sugar, says 'however obscure the mechanism may be we have reached a point at which we may say that the level of sugar in the blood is kept in balance by insulin which lowers it when it is too high, and the pituitary, adrenal cortex, and thyroid which raise it when it is too low. To this regulation, only one organ, the liver, is absolutely necessary. In health, the main regulator of the blood sugar is insulin, to which the liver constantly responds by increasing or decreasing its output of glucose from the glycogen in its stores. In this process the endocrines "set the temperature of the furnace". In sugar diabetes the "temperature" is set too high: in the presence of a tumour of the insulin-secreting cells of the pancreas it is set too low. When the adrenal cortex overworks it is set too high; when it is sluggish the setting is too low.' Exactly how this regulation is brought about, Greene writes, is still unknown in detail, but it may be that the various enzyme systems involved in sugar balance are affected by hormones. To this picture of sugar balance we must add the hypothalamus whose 'feeding' and 'satiety' centres we have talked about before in Chapter 17. These centres are sensitive to sugar, and if the hypothalamus is damaged by disease, levels of blood sugar can become too high or too low. The hypothalamus is again seen in a superior light, here on its control over sugar metabolism. Clearly it is remarkable that most of us balance on this chemical tight-rope all our lives, and are thankfully normal.

A BASIC PUZZLE

Another point of interest is, how do hormones do what they do? Here we have to go to the fine structure of the body, the cell. Enzymes, catalysts that drive the chemical reactions in the body cells, are almost certainly affected, as indicated above. At this *molecular* level, hormones (insulin, for instance) might play a key part in modifying the permeability of cell membranes. In the case of insulin, the barrier set up to sugar by the cell membrane in fat and muscle cells is reduced by the hormone allowing the cells to take up sugar from the blood.

The basic unit of heredity, the gene, might also be affected by hormones. As yet the strongest evidence for the regulation of gene action by hormones comes from insects, but indirect evidence has come from birds and mammals.

DNA (deoxyribonucleic acid) is the chemical that genes are made of and, most important, it is the *architect* of the proteins, the foundation-stone of life. RNA (ribonucleic acid) is the builder of the proteins, and works from the plan issued by the genetic architect DNA. Administration of hormones (thyroxin, insulin, pituitary growth hormone, oestrogen, and testosterone, for example) in mammals can cause not only an increase in the amount of RNA in the cell nucleus and cytoplasm but also the appearance of new kinds of protein. By contrast a hormone can stop DNA acting as the issuer of a blue print and so prevent protein synthesis.

It has been suggested, with little evidence, that a hormone might expose hidden stretches of DNA (a block of genes) to allow new kinds of protein to be synthesized by RNA or, on the other hand, to mask gene blocks to repress protein formation. All this detail has been given only to indicate that the finer mechanism of hormone action is one of the major challenges of physiology and biochemistry. For instance, how can a single hormone activate a group of genes which are functionally related (to produce, say, a given protein blue-print) yet physically separated? And how are the sets of genes activated by a given hormone 'chosen'? Does a hormone go to the chromosome direct and exert its effect 'in person'?

Of course we know now after over 200 years of patient medical and scientific advance what (some) hormones do, simply by studying the effects of animals and man of total destruction by surgery or disease of a particular gland, or by observing the effects of 'replacement therapy', that is making good from outside sources—extracts of glands taken from animals—what the body has lost. But how they do what they do is by no means clear.

EQUILIBRIUM

One remarkable aspect of the body, as we have noted, is that despite a fluctuating external (and internal) environment—heat, cold, wet, dry, altitude, and so forth, besides what we make our bodies do by sitting, standing, lying, running, eating, working, and so on—it maintains itself in equilibrium. Man, like other animals, has a great many devices by means of which his body keeps itself in this steady state, so that despite, for instance, a fluctuating external temperature, the blood temperature shifts only by about one degree celcius. In other words, the internal climate of the body remains steady; not only blood temperature, but the temperature of all the body fluids, their acidity, their oxygen and carbon dioxide concentration, their mineral content, their sugar supply (as we have seen), and so on. Physical systems, like blood pressure, are also kept within reasonable limits, but are engineered with good safety margins so that extra demands can easily be met.

The basic importance of the constancy of the 'internal environment' is the protection of the delicate living enzyme systems of the body whose chemical activities in every body cell keep us alive. Enzyme systems, which are partly protein, are easily destroyed by strong heat and acidity, and work best in quite specific and mild conditions of temperature and hydrogen ion concentration.

The endocrine system is not exempt from these regulating mechanisms, for its balanced activity is necessary to health. As we have seen, complicated mechanisms maintain the blood sugar within very narrow limits, presumably because the constancy would not be possible with simpler systems. Demand for sugar by the tissues sets in action knife-edged mechanisms that release sugar from reserves, stimulate its production, and activate animals, at least, to hunt for fresh supplies. All these devices enable sugar to be kept in balance. A failure of the mechanism in man results in diabetes. The best way to describe this maintenance of a steady state in spite of changes in the surroundings (homeostasis) is to compare living systems with machines. Indeed, this method has led to a clearer view of the concept of control and has made possible a language for describing control of the body.

SERVO-MECHANISMS

A device in operation both in machines and in the endocrine and other systems of the body is called a servo-mechanism or 'slave' mechanism. In a mechanical device it enables a distant operator to control a powerful engine with precision yet with a light touch. The steering system of a large ship is a good example. Three features are essential in a servo-mechanism and we can see these in action in the endocrine company, taking the thyroid-pituitary-hypothalamus axis as an example.

1. There needs to be a coded pattern of instruction for the machine: here the information of heredity laying down a blue print for the controlled release of thyroxin under appropriate conditions, different for different people.
2. There needs to be a receptor or receptors to record the state of the output system: here detectors in the anterior pituitary gland, also neurones in a hypothalamic region, both of which monitor the output of thyroxin from the thyroid.
3. There needs to be a comparator to measure the difference between the instructions and the course the machine actually follows and to provide a controlling output that reduces any mismatch: in our example chemical signals (thyroxin) circulating in the blood inform the detectors ((2) above) which act as comparators as well as detectors of the quantity of thyroxin circulating. A fall in the amount of thyroxin stimulates the release of TSH (thyroid stimulating hormone) from the anterior pituitary. When the concentration of thyroxin rises above a certain level, the secretion

of TSH is inhibited. TSH, the chemical signal sent from the anterior pituitary, can only be decoded by organs sensitive to it—in this case, the thyroid.

Another term needs to be introduced here, called by engineers 'feed-back', meaning return information of the effects of any controlling action. Thus to continue our example: thyroxin feed-back alters the sensitivity of neurones (detectors) in the hypothalamus which control the releasing factor output which triggers TSH. (Feed-back from the thyroid may, however, affect TSH production by the anterior pituitary direct.)

Similar patterns of control exist between other sets of endocrine organs. For example, the anterior pituitary produces a hormone, ACTH, which stimulates the adrenal cortex to secrete its steroid hormones. These steroid hormones help to regulate carbohydrate metabolism and the balance of sodium in the body fluids, and are parts of a servo-mechanism. If one adrenal gland is removed, the ACTH stimulation causes the cortex of the remaining adrenal to increase in size. If, on the other hand, the anterior pituitary lobe is removed, the resultant lack of ACTH leads to a shrinkage of the cortex of both adrenal glands. If ACTH is then injected, the adrenals return to their usual size. It appears that under normal conditions the concentration of adrenal cortex hormones in the blood controls the secretion of stimulating ACTH by the pituitary: when this concentration is high, less ACTH is released; when it is low, the release of ACTH increases. A similar balancing mechanism is believed to operate in other cases of hormones stimulating a specific target organ. The secretion of sex hormones by the ovaries and testes is subject to a similar feed-back control, as we shall see in the next chapter, and the various sex hormones in turn interact with each other. Thus the effects of oestradiol, one of the two hormones produced by the ovaries, may be neutralized by an extra output of progesterone, the other ovarian hormone. There is, however, considerable ignorance about how all these sets of interactions are organized into a pattern.

Interestingly enough, in the endocrine system there are wheels within wheels; besides the large feed-back loops already described, there is also a short 'internal' feed-back circuit between pituitary and hypothalamus whereby the pituitary hormones regulate their own 'releasing' factors which we have spoken about earlier. Blood containing active substances released from the pituitary can flow back to the region in the hypothalamus where releasing factors enter the blood stream.

Clearly the complexity of the system makes for great adaptability and delicacy of control. The control is so marvellous that in health, the 'pull and push' mechanisms involved never distort one way or the other before equilibrium is restored again.

CHEMICAL DECISIONS BEFORE BIRTH

This balance and delicate chemical control begins long before birth. Knowledge of hormone action before birth is still meagre, but two pieces of evidence—on the development of sex and the development of the cerebral cortex—give us a glimpse of the chemical machinery set in motion by the chromosomes at about six weeks after conception. Sex is decided at conception by the X-Y chromosome mechanism described in Chapter 1. Individuals with an XX chromosome complement develop into females, XY individuals into males. The basic decision—male/female—is dependent almost exclusively on the Y chromosome. With no Y (just one X), a female develops; with a Y, a male. But there is an enormous gap. What do the sex chromosomes actually do to cause sex development?

A theory based on experimental work with rats and with rabbit embryos suggest that a large piece of the sex chromosome determines the *rate* at which the uncommitted primitive sex organ (gonad) grows in mammals and possibly man: quick growth gives a boy, slower growth a girl. To explain this apparently simple recipe a little further: for the first five or six weeks after conception we are apparently sexless except, of course, for the XX, XY component in the chromosome complement. In about the seventh or eighth week the sex organs develop from a primitive structure associated with the developing kidneys, and we become 'double-sexed' or hermaphrodite. This primitive structure consists of an outer rind of cells, the 'cortex', and an inner core, or 'medulla'. (See Figure 16.) The cortex has the potential for developing ovaries and the medulla the potential for the testis. If a Y chromosome is present the medulla continues development, the the cortex dwindles, and the gonad becomes a testis. In a female (XX) the cortex of the primitive gonad develops into the ovary and the inner medulla regresses.

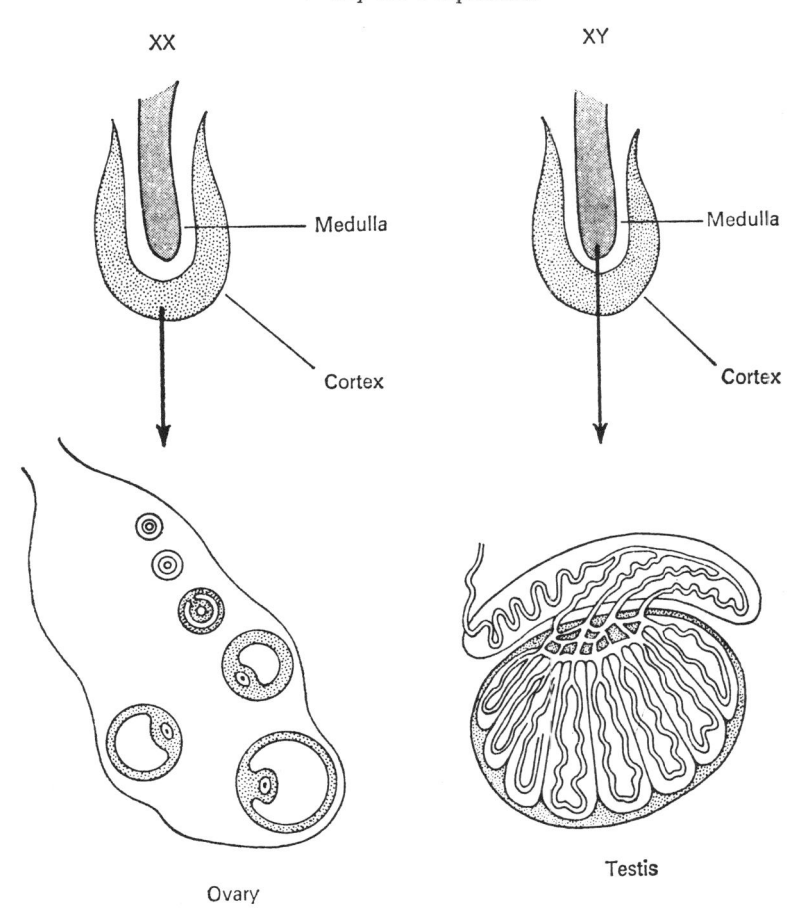

FIGURE 16 The development of the ovary (above) and testis from the early uncommitted sex organ in mammals. In individuals with a female (XX) chromosome make-up, the outer cortex of the primitive gonad develops into the ovary, while the inner medulla regresses. The opposite pattern occurs in XY individuals, where the testis grows from the medulla at the expense of the cortex. (From Mittwoch, 1971.)

MAN IS CARVED OUT OF WOMAN

Some interesting experiments by A. D. Jost in Paris on rabbit embryos shows that male development takes place only if testes are present in the embryo. If the testes are removed from young embryos, female development goes on despite a male (XY) chromosome complement. Normally, of course, a Y chromosome causes testes to develop and they provide the male hormone, testosterone, which programmes the embryo to continue development as a male. If there is no Y chromosome, an ovary forms. The ovary in the embryo does not appear to secrete any sex hormones, but female development goes on. On the basis of this evidence feminization is apparently due to:

1. *absence* of a Y chromosome, and
2. consequent *absence* of male sex hormone and not to the *presence* of oestrogens.

If a Y chromosome is present, with normal male development and testosterone acting, how might this arrangement affect the development of the primitive gonad into a male? V. Mittwoch has shown that the gonads of male embryos in the rat are larger than the gonads of their female litter mates long before testes or ovaries are detectable under the microscope. The Y chromosome, then, might step up, perhaps through testosterone, the growth of the developing gonad by faster cell division (mitosis). Perhaps, Mittwoch argues, a gonad can become a testis only if it reaches a certain size before a given stage in development. If it does not reach this critical size, the gonad will develop into an ovary. Eve was made from a rib of Adam, so the Old Testament says, but it seems from the above evidence that maleness is imposed on a basically female pattern. This notion is supported further by experiments on rats which show that sexual behaviour patterns are fundamentally female, and that male patterns are induced by the direct action of testosterone on the hypothalamus in the critical first few days after birth. The brain must then differentiate into male and female types so that permanent control over the activity of the pituitary is established and male and female sexual behaviour initiated. How testosterone imposes its sex-difference imprint on the hypothalamus is unexplained.

We must be careful to underline the fact that these are the results of experiments on rats and rabbits, but it is not unreasonable to suppose that the human brain differentiates in a similar way in the womb under the action of testosterone. Indeed, they have permitted new interpretation of many abnormalities of the human genital organs. For instance, it was predicted that people afflicted with total absence of sex glands from birth would have a feminine shape no matter what their genetic sex was. This was confirmed some ten years ago when the appropriate studies of sex chromosomes became possible. It was then discovered that absence of sex organs and feminine genital tract is often linked with absence of one X chromosome (XO). The unfortunate people with this syndrome have underdeveloped female bodies and vestigial infertile sex glands. (See Chapter 13.)

Of course, the fundamental difference between sexual behaviour of rats and rabbits and that of man is that the animals' behaviour depends largely on hormones circulating in the blood stream acting on the brain. In man, as we keep emphasizing, the cerebral cortex plays a larger part in sexual behaviour than gonadal hormones, but he is not entirely free from their influence, as we shall see.

THYROID HORMONE AND LEARNING ABILITY

A gland that is closely associated in the layman's mind with mental changes is the thyroid, which we shall return to in the next chapter. Thyroxin, a thyroid hormone, is necessary for the development of the brain. When the thyroid gland has partially or wholly failed to develop during foetal life, or where iodine, a raw material of thyroxin, has been in short supply so that the gland cannot work properly, mental development is severely retarded and the infant will become permanently cretinous (that is severely mentally retarded with labile and uninhibited emotions if treatment is not prompt after birth.

What might be the basic cause of this serious disability? The evidence in the main comes from rats, where comparisons can be made between the brains of normal infant rats and those which have had their thyroid gland removed immediately after birth. These latter 'cretinoid' rats show deviations from the normal in the neurones of the cortex: the 'wiring' is not so complex, that is the density of axons and dendrites is lower than normal; the dendrites branch less and are shorter, and the development of the axons down which impulses are fired from one nerve to the next is impaired. As a result, the animal's reflex responses are slower than normal. Some cells of the cortex suffer more from thyroid lack than others; those in 'layer four', for example, are associated with the outflow of nerves from the thalamus which in man might give emotional tone to thoughts which otherwise would be colourless. Lack of thyroid hormone, then, seems in rats to interfere with normal growth of neurones and consequent behaviour (learning ability and alertness), perhaps through the inability of the nerve cells to absorb nutrients from the blood necessary for protein synthesis, the enzymes necessary for the process being absent. Here, then, a fault in the chemical machinery *before birth* in humans can cause grave and often irreversible results on the mind and personality. Do the hormones of the endocrine system other than thyroxin and testosterone exert differentiating effects on the developing brain? If testosterone at a critical period of development does produce sexual differentiation in the hypothalamus, by what mechanism does it do so? And to what extent may hormones acting on the brain during infancy shape the future personality of an individual? In the case of thyroxin there is some evidence. The next half-century of research may well tell us much more about these knife-edged chemical operations which work before and after birth to influence personality.

19: The Chemistry of Character

GRAINS OF TRUTH

NAPOLEON, a small and very energetic man, started to grow fat around the age of forty. Not only did his body begin to change from statuesque slimness, but the alert, quick-thinking mind, which digested information rapidly and paid meticulous attention to detail, became slowly more sluggish and indecisive. Over the years till his death at fifty-one in 1821, his body began to take on an effeminate look, hairless and soft-skinned, and his virility waned. Though obviously speculative, some endocrinologists have put down these changes, which started just before his disastrous Russian campaign, to pituitary malfunction caused by Fröhlich's disease. It does seem that after about 1809, when the physical and mental changes started, the tide of success turned. Perhaps Waterloo might have been won by Napoleon had his health been better. On the other hand, the pressures on his life would be a reasonable basis for fatigue without bringing his glands into the picture. Post-mortem findings by an English physician indicate a picture of pituitary malfunction.

'The whole surface of the body was deeply covered with fat. Over the sternum, where generally the bone is very superficial, the fat was upwards of an inch deep, and an inch and a half to two inches on the abdomen. There was scarcely a hair on the body, and that on the head was fine and silky. The whole genital system (very small) seemed to exhibit a physical cause for the absence of sexual desire and the chastity which had been stated to characterize the deceased during his stay at St Helena. The skin was noticed to be very white and delicate, as were the hands and arms. Indeed the whole body was slender and effeminate.'

Besides this, Napoleon was dogged with other ill health. The post-mortem found a large stomach cancer, something his father had died from; he suffered from bladder trouble and from agonizing piles. When Wellington's army was in a very vulnerable position at Quatre Bras, Napoleon, perhaps because of an incapacitating attack of inflamed and prolapsed piles, delayed his attack and gave Wellington a precious day to move his troops to the more favourable terrain of Waterloo. It was 'a damned nice thing, the nearest run thing you ever saw in your life', as Wellington probably said, but whether it was piles or pituitary failure or the sapping effect of a cancer, Waterloo was lost by Napoleon.

G. I. Cobb, in his analysis of Henry VIII, is of the opinion that the royal machine was controlled on the mental side by a sense of inferiority and on the endocrinological side by an 'active thyroid-pituitary' combination. Towards the end of Henry's life, so it is postulated, his thyroid began to give out. He grew fat, puffy under the eyes, and, as shown by Holbein's portrait, lost the outer halves of his eyebrows, regarded as a characteristic of under-functioning of the thyroid gland. The historical facts given by Scarisbrick, and borne out by portraits, show that Henry's bullock body was a magnificent piece of nature's handiwork which served him well for the first thirty-five years of life, and his mind in youth was acute. His body measurements at twenty-three were: chest 42 inches (107 centimetres); waist 35 inches (89 centimetres). In his forties he put on weight and in the 1540s became the tortuous and moody hulk who had to be hauled upstairs by machinery. His measurements (as determined by his armour) at 50 were: chest 57 inches (145 centimetres); waist 54 inches (137 centimetres). Apart from this gross corpulence he was plagued with leg ulcer. Perhaps his deterioration was indeed due to endocrine failure. The engraving by the Flemish engraver Cornslys Matsys done towards the end of his life in 1548 shows the swelling of the eyelids to give the pig-like look of thyroid deficiency (myxoedema). Perhaps Henry's irritability and moodiness in later life were due to the frustration of an acute mind which realizes its new incompetence over which it has no control. A drop in thyroid activity can have this effect. But there are other hypotheses about Henry's health. He might have suffered from a metabolic disease—gout, inherited from his father Henry VII. Aggravated by gross feeding, gout can trail a host of ills that could explain Henry's personality changes and physical ills. Gout is not a killer but its associated conditions are: arterial degeneration; nephritis; thrombosis and embolism. Shrewsbury, whose hunch this is about gout, writes:

> Superimpose upon his basic egotism with its self-love, its self-will, its vanity and its insensate jealousies, the rising blood pressure, the pain, the depression, the irascibility and the mortification that are the concomitants of gout, and it seems to me that we have an adequate explanation for the change from the handsome, athletic, magnificent young 'Prince of the Renaissance', the spoilt darling of the English parliament and people, to the suspicious, jealous, lonely old despot.

These descriptions of how endocrine or other metabolic disease can affect great historical personalities are amusing but are bound to be based on bits and pieces of evidence, and can wander away from science into science fiction. It is too simple anyway to assume that a type of personality which goes, for instance, with a gland disturbance, is the direct result of that disturbance. A pituitary dwarf may be cocky or aggressive, but his cheek may be due to the unsympathetic stares and whispers of people rather than to pituitary malfunction. Brought up among pygmies, his personality might be different. As for a change in personality due to endocrine disease, there appears to be some truth in this in some diseases, but the effect of disease is not, it seems, to *remake* the personality but to influence what is there already. Those who walk a mental tight-rope are easily unbalanced as a result of an excess or deficiency of hormone. Thyroxin circulating in the blood, in excess, notches up the reactivity of the autonomic nervous system and the brain, making the sufferer more irritable, anxious, and excitable than he was before. By contrast, lowered thyroxin slows down his emotions. In the former case an excitable personality becomes more excitable, in the latter calmer, but according to the fineness of mind, more or less frustrated. The relationship between genes, hormones and the human mind and nervous system is a web the complexity of which we have hardly begun to understand, but modern evidence has made a start.

FIRMER EVIDENCE FROM THE SEX GLANDS

The sex hormones perhaps give the clearest indication of a connected and complex web, rather than a chain, between chromosomes, endocrines, brain, and psychology. As we have seen already, there is good evidence of the Y chromosome as the male determiner, a decision made at conception. The general femaleness of a Turner's syndrome subject (XO, the Y chromosome having been lost) shows that the Y is strongly male-determining, and this view is reinforced by another unfortunate abnormality, Klinefelter's syndrome (XXY). Despite two X chromosomes, the Y overrides their influence to produce a nearly normal male. We know, too, from mammalian experiments, that the presence of a Y in an XY normal male constitution, triggers an early and rapid development of the testis in the primitive sex organ long before birth, which in the presence of XX (normal female) would take the slower course of development characteristic of the ovary. In other words maleness overrides femaleness. We know again, from mammalian experiments, that animals, and perhaps man, are born with a sex difference imprinted somehow on the hypothalamus by testosterone. This imprint determines the pattern of pituitary effects on the sex organs: cyclic in the female and non-cyclic in the male (see Figure 17) and we shall consider this further in a moment. We know, too, that the X and Y chromosomes act as a trigger which channels development in two paths but the paths are broadened, as we have suggested already, by the variation in all forty-six chromosomes, not just the Xs and Ys, to give variations in the secondary sexual characters, size of the genitals, and in the types of intelligence and temperament. Thus some men of normal (XY) constitution have 'female' temperaments and intelligence, and some women of normal genetic sex (XX) have a 'masculine' temperament and intelligence. Complicating the web of relationships is the part played by the adrenal glands in sex 'drive'. In women the androgens released from the adrenals may play a larger part in sex drive and desire than the ovarian hormones which probably maintain the female genitalia functionally and so expedite coitus, although the ovarian hormones might make the

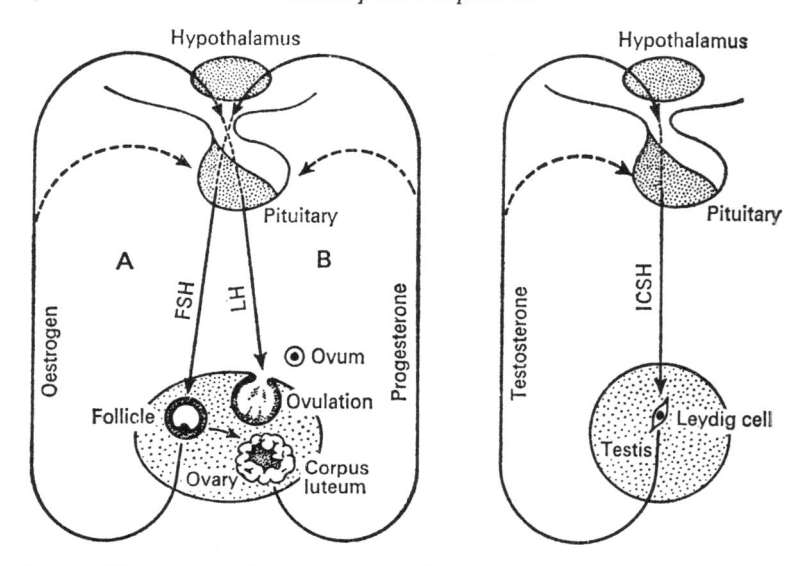

FIGURE 17 Interplay of sex hormones differs in the female mammal (*left*) and the male (*right*). In the cyclic female system the pituitary initially releases a follicle-stimulating hormone (FSH) that makes the ovary produce oestrogen (*arrows at A*): the oestrogen then acts on the hypothalamus of the brain to inhibit the further release of FSH by the pituitary and to stimulate the release of a luteinizing hormone (LH) instead. This hormone both triggers ovulation and makes the ovary produce a second hormone, progesterone (*arrows at B*). On reaching the hypothalamus the latter hormone inhibits further pituitary release of LH, thereby completing the cycle. In the non-cyclic male system, the pituitary continually releases an interstitial-cell-stimulating hormone (ICSH) that makes the testes produce testosterone; the latter hormone acts on the hypothalamus to stimulate further release of ICSH by the pituitary. Broken arrows represent the earlier theory that the sex hormones from ovaries and testes stimulated the pituitary directly. (From Levine, 1966.)

woman more sexually inclined at certain times of the menstrual cycle—mid-cycle, which is the time of ovulation. Loss of ovaries and their hormones seems to have no effect on sex drive in some women, but loss of adrenals cuts out drive, activity, and response.

In men, Greene writes, the male hormone, testosterone (the most important of the androgens) has only a slight influence on potency. Perhaps the key factor is the way testosterone imposes itself on the mind in foetal life and early childhood so that sexual desire is developed. The male castrated in youth, like one born with testicular deficiency, rarely develops desire, but the man castrated in later life is not necessarily lacking in desire or in the capacity to have intercourse. It is tempting to suggest that the flood of testosterone and ovarian hormones at puberty activates not only physiological machinery such as menstrual cycles, but also patterns of sexual behaviour laid down in the womb.

Homosexuality

A point on homosexuality in men and women is relevant here. A great deal has been written about homosexuality, and research has been unable to detect any physical (hormonal) abnormalities to account for it. Environmental circumstances and relationship problems have been put down as the roots of a psychological condition: a puritanical upbringing; traumatic early experience with the opposite sex; a dominant mother; a weak, absent, or cruel father; and lack of confidence in normal relationships with the opposite sex, have all been separately or jointly blamed. C. D. Darlington suggests that about 5 per cent of men and women are born with a homosexual tendency. Among kings, William Rufus, Richard Coeur de Lion and Frederick the Great, were thought to be homosexual. All failed to beget children, perhaps because of the strength of the homosexual drive.

Now there is some evidence for hormonal causes underlying at least some homosexuality in both men and women. J. A. Loraine has recently discovered abnormal sex hormone levels in a small sample of homosexual people.

This work was carried out with homosexual volunteers of both sexes and compared with results from a larger control group of normal heterosexual men and women. Twenty-four-hour urine samples were checked continuously over periods of nineteen to twenty-six days, and the results showed levels of the male sex hormones, testosterone, to be generally below the normal range, in the homosexual men.

In the lesbian women, not only was the female sex hormone, oestrogen, below the normal range, but tests also showed significantly higher levels of male testosterone than in the control group of normal women.

Although the sample was small, this was offset by the duration of the tests, and since publication more evidence of abnormal homosexual chemistry has come from Margolese, who has also found reversal of male hormone proportions in homosexual as opposed to heterosexual subjects.

Loraine suggests that the importance of such findings does not lie in what they reveal about the existing hormone balance. Rather it is the inference of corresponding hormone patterns in these people *before* birth, acting during the vital intra-uterine stage, when the hypothalamus of the foetus is known to be so vulnerable to hormone influence that it imprints wrongly to give an abnormal pattern of behaviour.

Perhaps, too, the early influence of hormones on the brain influences other sexual abnormalities. Some of the most tragic of these are the male and female transexualists who believe they belong to the opposite sex yet are trapped in the wrong body and wrong gender. Armstrong believes (without direct proof at present) that recent work on sexual differentiation of the brain might eventually provide an answer. Work with monkeys has already shown that if sufficient androgens are not available in the foetus at the critical time, the brain develops the female pattern. Armstrong suggests that a similar process in humans could account for a male body with a female-pattern brain. The amount of androgen required for normal masculinizing of the brain is greater than that required to masculinize the genitalia, and the critical stage for brain differentiation is later, so that sufficient androgen might be present at the right time to produce an anatomically normal male in respect to other criteria of sex, but still leave the brain not properly differentiated.

The nervous system too, is woven strongly into this web of sexual behaviour. Erection of the penis is under nervous (parasympathetic) control, and anxiety or fear is a common cause of impotence, though testosterone appears to be important in the maintenance of an erection. Of course the androgens, testosterone in particular, are of profound importance in growth and the development of the male secondary sexual characters at puberty: the extension of pubic hair and growth of beard and moustache; the deepening of the voice and the aggressive attitudes of young males are caused by a flood of testosterone on the immature brain.

At the hub of this complex web is the human mind with its capacity for consciousness, its memory stores, its intelligence, closely enmeshed with the emotional centres, as we have seen. Because of the brain, sexual desire in man is less dependent on, and much more variably determined by, hormones than in lower animals. In male rats with severe brain damage, testosterone can repair sex interest, impossible in man. Destruction of the whole cortex in female rats will not stop the mating response.

From what we now know about genetics, each person has a unique genotype, a unique sexual character attracted by different qualities. In Chapter 2 we showed that marriages tend to be 'assortative'; that is they involve the discovery of other individuals who are alike enough in mind and character for us to be able to live and work with. Brain, hormones, nervous system, and genes, then, interact to create different sexual patterns and behaviours. The complexity and delicacy of the checks and balances in this sexual aspect of the human machine are truly wonderful.

STEROIDS AND THE MIND

The waxing and waning tides of steroid hormones in the menstrual cycle, in pregnancy, and after child birth, can affect the mind profoundly. Figure 18 summarizes the normal cycle. Briefly, what happens is that at puberty a signal, about which nothing is known, causes the hypothalamus to trigger the release from the pituitary of FSH and LH, both of which act on structures in the testes of boys and the ovaries of girls to bring about sexual maturity. In women of child-bearing age, waxing and waning tides of FSH and LH each month act on the ovarian follicles and cause them to ovulate (release the egg). Under the influence of FSH a ripening follicle becomes a miniature endocrine gland which secretes oestradiol. This stimulates the regrowth of the lining of the womb (stripped away at the previous menstrual period) for possible pregnancy, causes the breasts to grow in preparation for lactation, makes the vaginal walls more elastic, and softens the plug of mucus in the cervix. In addition it causes a retention of water in the tissues at the time of ovulation and just before menstruation. A few women gain as much as twenty pounds in weight, most about two. Up to day fourteen in the twenty-eight-day cycle FSH is in the ascendant, but its very action—to cause increasing amounts of oestradiol—rapidly cuts down the secretion of FSH, by oestradiol feeding back to the hypothalamus which in turn acts on the pituitary to cause the waning in FSH. At the same time LH is on the ascendant, and when the balance between it and FSH becomes critical, ovulation takes place; the egg is discharged from the follicle towards the Fallopian

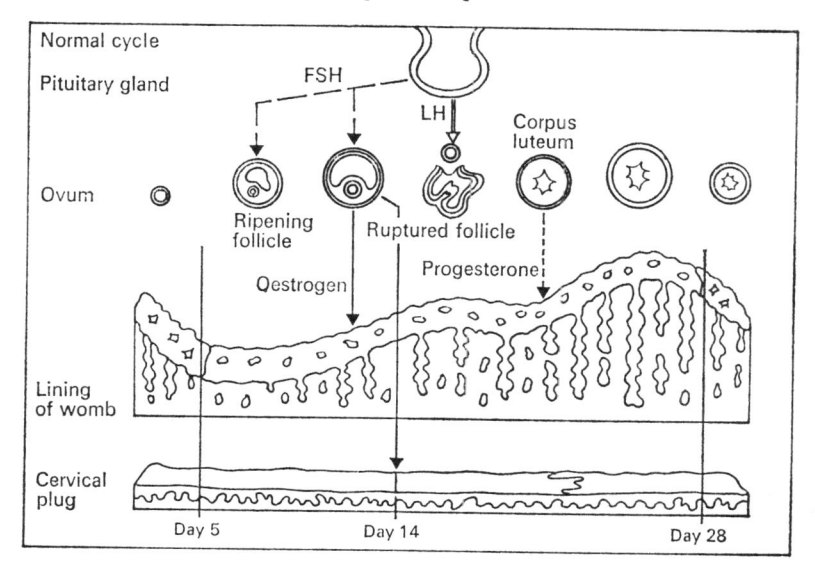

FIGURE 18 Ovaries contain hundreds of thousands of unripe sacs or follicles, each of which contains a potential ovum, or egg. At the beginning of the monthly cycle, while bleeding is taking place, several are maturing under the influence of the follicle-stimulating hormone, or FSH, produced by the pituitary. FSH is released only when very little of the hormone oestrogen is being produced by the ovaries.

The maturing follicles secrete fluid in their cavities. This fluid contains oestrogen, which 'primes' the lining of the womb again to prepare for a possible baby. It seems also to thin out the plug of mucus which normally blocks the cervix, thus making it possible for sperms to enter the womb.

The increasing amounts of oestrogen rapidly diminish secretion of the follicle-stimulating hormone (FSH) by the pituitary gland. At the same time a second pituitary hormone is released in greater quantities—the luteinizing hormone, or LH. This will make ovulation possible. About fourteen days after the first day of menstruation, when the balance between FSH and LH is critical, one of the follicles, which has grown more than the others, ruptures. It discharges its egg towards the Fallopian tube leading to the uterus. This is what is meant by ovulation.

After losing its egg, the ruptured follicle changes into a solid structure—the yellow body, or corpus luteum. This corpus luteum continues to produce oestrogen, as its parent follicle had done. It also produces the second ovarian hormone, progesterone, which acts on the pituitary to stop production of LH. Without LH and FSH no other follicles can mature fully (they, and their eggs, tube leading to the uterus. There is some evidence that the ovary itself informs the hypothalamus and pituitary of its state of readiness for ovulation by oestradiol feed-back. Put crudely, the pituitary-hypothalamus combination becomes the slave of the ovary which says 'ovulate me please' and this signal causes a rise in LH and subsequent ovulation.

Once a follicle has ovulated, it becomes a temporary gland, called the *corpus luteum* under the influence of LH, and secretes progesterone which completes the lining of the womb in conjunction with oestradiol which continues to be secreted. The role of progesterone is to cause a secretory change in the regenerated lining, presumably to provide food for the fertilized egg before implantation. Progesterone in turn feeds back to the hypothalamus and LH secretion is stopped. Without LH and FSH, no other follicles can mature fully. About day twenty-four, if no fertilized egg has arrived in the womb, the *corpus luteum* atrophies and its production of oestradiol and progesterone is at a low ebb. The egg left lying in the womb is discharged and the lining of the womb collapses, and on day twenty-eight menstrual bleeding starts. Once oestradiol production is down. FSH is once more released and a new crop of follicles begins to ripen.

If an egg is fertilized, the placenta, the structure through which the developing foetus is fed, secretes a hormone which stimulates the secretion of progesterone and oestradiol by the *corpus luteum* of the ovary. After about three months, the placenta itself takes over the production of oestrogens and progesterone for the remainder of pregnancy. The relatively large amounts of these two hormones prevent the pituitary from sending out FSH and LH needed for ovulation. Perhaps the foetus itself participates in triggering the birth process, through the production by its pituitary gland of ACTH which activates the foetal adrenal gland to produce a steroid hormone which, acting through the placenta, stimulates the uterine muscles to contract, leading to birth. In addition, this hormone probably prepares the foetal lungs for breathing.

---

degenerate and wither away). Progesterone also completes the preparation of the lining of the womb for the reception of a fertilized ovum.

Around day 24, if no fertilized ovum has arrived in the womb, the corpus luteum atrophies like the other ripe follicles of that cycle and its production of oestrogen and progesterone ceases. The lining of the womb then collapses, and on day 28 menstrual bleeding begins. Once oestrogen production is down, the pituitary gland can begin to release the follicle-stimulating hormone once more, and a new crop of follicles begins to ripen.

THE INJUSTICE OF HORMONES

How does this chemical machinery affect the brain? Premenstrual distress during the fortnight or ten days before 'the period' is common: about a third of women suffer. Some just feel irritable, others less fortunate suffer considerable mental distress, and in this condition the body weight increases through water retention. Some women in this state feel tense, depressed, restless, emotionally unstable, mentally dull, and liable to weeping attacks and personality changes.

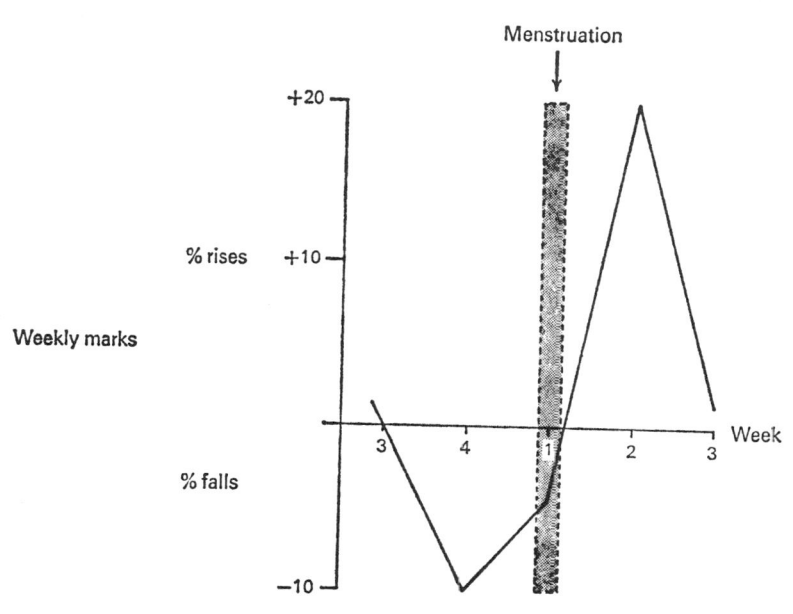

FIGURE 19 Variation in schoolgirls' marks due to menstruation. (From Dalton, B.M.J., 1960.)

Among schoolgirls, school work suffers at this premenstrual stage, and discipline troubles and untidiness are more frequent. Some girls' examination results are poorer than expected and prefects hand out more punishments. K. Dalton, in one English study of 217 menstruating girls between the ages of eleven and seventeen, showed that one-quarter had a fall in weekly marks during the premenstrual stage followed by a rise after menstruation (see Figure 19). The marks during the menstrual week reflect the failure to learn grammatical rules in

English, mathematical principles, and French and Latin, during the premenstrual week, and also the improvement of scores as water retention is relieved accompanying the full menstrual flow. Though school tests impose a stress, the strain of full-blown examinations increases premenstrual symptoms and imposes a handicap. About one in six girls in any examination entry will be in her premenstruum and thus at her lowest intellectual ebb. As Dalton says, 'while zealots campaign assiduously for equality of the sexes, nature refuses to grant equality even in one sex.'

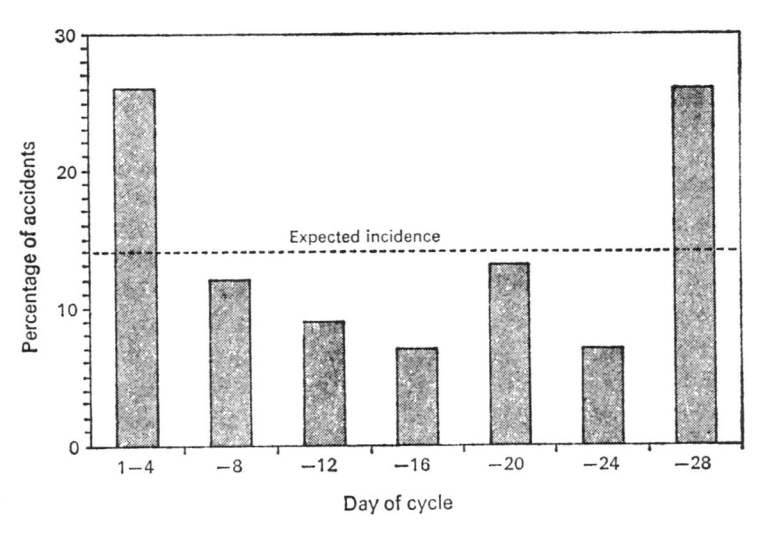

FIGURE 20 Distribution of eighty-four accidents in the menstrual cycle. (From Dalton, B.M.J., 1960.)

A higher proportion of crimes are committed in the premenstrual or early menstrual phase, and suicides, accidents (see Figure 20), and deaths from disease are commoner.

In one American study, half the female prison population suffered from premenstrual tension, and 60 per cent of crimes (murder, manslaughter, and assault) were committed in the premenstrual week and 17 per cent during menstruation. A similar study in Paris showed that 84 per cent of all the crimes of violence by women were committed at the premenstrual or early menstrual phases of the cycle.

Probably the alterations in the mental state depend somehow on the oestradiol-progesterone balance, but the evidence is not clear. There is some evidence that low blood sugar is a feature of premenstrual tension and we know that sugar level affects the mind: low blood sugar causing anxiety, irritability, and tension in one person, and violence in another.

Perhaps, too, progesterone formed by the placenta during pregnancy causes the placidity of pregnant women. The risk of suicide and mental illness is low in pregnancy, but greater than normal a few days after birth. Perhaps this mental disturbance is due to a sudden withdrawal of hormones which have been in high concentration during pregnancy. Clearly the mental effect of these steroid hormones is powerful, but again it is likely that the 'set' of the personality is important: some women are more easily pushed into crime, suicide, accident, or quarrelling by steroid imbalance, others remain on an even keel.

This brief account of hormones shows that although they do affect growth, temperament, and intelligence, they are but one ingredient in an integrated pattern. The hormones react with the structural foundations in every organ, but particularly with the brain, to influence behaviour, while the 'setting' of the brain for sex difference and intelligence and even temperament, might happen before birth under the control of foetal hormones. Chromosomes and genes do affect hormones, brain, nervous system, as we have seen in earlier chapters, but we are only at the beginning of understanding how they *interact* to influence man and his personality. Perhaps in the next century this will become clearer, but whether we shall be wiser or happier is another matter.

Bibliography

Some books and papers consulted in compiling this book. An asterisk indicates books which are suitable for the general reader.

PART 1 THE ROOTS OF PERSONALITY

*Bodmer, W. F. and Cavalli, L. L. (1970) 'Intelligence and Race', *Scientific American*, October.
*Brierley, J. K. (1967) *Biology and the Social Crisis*, Heinemann.
Burns, M. and Fraser, M. (1966) *Genetics of the Dog*, Oliver and Boyd.
Burt, C. (1961) 'Intelligence and Social Mobility', *British Journal of Statistical Psychology*, **14**.
—— (1959) 'General Ability and Special Aptitudes', *Educational Research* **1**.
—— (1959) 'Class Differences in General Intelligence', *British Journal of Statistical Psychology*, **12**.
—— (1961) 'The Gifted Child', *Yearbook of Education*, Evans.
—— (1966) 'The Genetic Determination of Differences in Intelligence: a Study of Monozygotic Twins Reared Together and Apart', *British Journal of Psychology*, **57**.
Casey, M. D. *et al.* (1971) 'Patients with Chromosomal Abnormality in Two Special Hospitals', *Special Hospital Research Reports*.

Coates S. et al. (1957) *Arch. Dis. Childh*, **32**.
Cox, C. M. (1925–1926) *Genetic Studies of Genius*, Harrap.
Dalton, K. (1968) 'Ante-natal progesterone and intelligence', *British Journal of Psychiatry*, **114**.
*Darlington, C. D. (1966) *Genetics and Man*, Pelican.
—— (1963) 'Psychology, Genetics and the Process of History', *British Journal of Psychology*, **54**.
—— (1954) *Heredity and Environment*, Caryologia.
*Eysenck, H. J. (1964) 'The Biological Basis of Criminal Behaviour', *Advancement of Science*.
*—— (1964) *Crime and Personality*, Routledge.
Fraser Roberts, J. A. (1952) 'The Genetics of Mental Deficiency', *Eugenics Review*, **XLIV**.
Fuller, J. L. and Thompson, W. R. (1967) *Behaviour Genetics*, Wiley.
*Galton F. (1962) *Hereditary Genius*, Fontana.
Gray Jeffrey (1971) 'The biology of sex differences', *The Times*, 11 December.
Haldane, J. B. S. (1963) *Man and His Future*, Churchill.
Heim, A. (1970) *Intelligence and Personality*, Pelican.
Jacobs, P. et al. (1971) 'Chromosome Surveys in Penal Institutions', *Journal of Medical Genetics*, **8**.
*Jeeves, M. A. (1967) 'Scientific Psychology and Christian Belief', *Inter-Varsity Fellowship—Harvard Educational Review*, **39**.
Jensen, A. R. (1969): 'How Much Can we Boost IQ and Scholastic Achievement?', *Harvard Educational Review*, **38**.
Jervis, G. A. (1954) 'Phenylketonuria' *Res. Pub. A. Nerv. and Ment. Dis.* **33**.
Kellog, R. et al. (1965) 'Form similarity . . .', *Nature*, 11 December.
*Lange, J. (1931) *Crime as Destiny*, Allen and Unwin.
Lederberg, J. (1970) 'Racial Alienation and Intelligence', *Harvard Educational Review*, **39**.
McWhirter, K. (1967) 'Gaols and Genes', *Daily Telegraph*, March.
Money, J. (1970) 'Pituitary-adrenal and related syndromes of childhood: effects on IQ and learning', *Progress in Brain Research*, **32**.
—— (1971) 'Pre-natal hormones and intelligence: a possible relationship', *Impact of Science on Society*, **XXI**, No. 4.
*Parry, M. (1968) *Aggression on the Road*, Tavistock.
*Penrose L. S. (1959) *Outline of Human Genetics*, Heinemann.
Price, W. H. and Jacobs, P. (1970) 'The 47 XYY Male with Special Reference to Behaviour', *Seminars in Psychiatry*, **2**.
Shields, J. (1967) 'Twin Research and Psychiatry', *World Medical Journal*, **14**.
*Silcock, B. (1970) 'Race, Class and Brains', *Sunday Times*, 1 and 8 February.
Slater, Eliot and Shields, J. (1967) 'Genetical Aspects of Anxiety', *British Journal Psychiatry*, **3**.
Terman, L. M. et al. (1925 and 1947) *Genetic Studies of Genius*, Stanford.

*Thoday, J. M. (1965) 'Geneticism and Environmentalism' in *Biological Aspects of Social Problems*, Oliver and Boyd.
*—— (1971) 'Genetics and Educability', *Institute of Biology Symposium*, London.
Watts, K. P. (1953) 'Influences Affecting the Results of a Test of High-grade Intelligence', *British Journal of Psychology*, **XLIV**.

PART 2 A VIEW OF THE BRAIN

*Boyne, Alan (1971) 'The Shadow of Brain Research', *New Scientist*.
Broadbent, D. E. (1970) 'Psychological Aspects of Short-term and Long-term Memory', *Proc. Royal Soc. London*, **B 175**.
Close, H. G. *et al.* (1968) 'Sex Chromosome Anomalies', *Cytogenetics* **7**.
Darlington, C. D. (1970) 'Race, Class and Culture' (unpublished).
—— (1970) Review of *The Other Love* by H. Montgomery Hyde, in *Heredity*, **25**.
*Davy, John (1971) 'Memory', *The Observer*, March.
*Eccles, J. (1965) *The Brain and the Unity of Conscious Experience*, Cambridge University Press.
Eccles, J. (editor) (1965): *Brain and Conscious Experience*, Springer-Verlag.
*Elliott, H. C. (1970) *The Shape of Intelligence*, Allen and Unwin.
Fantz, C. D. (1961) 'The Origin of Form Perception', *Scientific American*, **204**.
Gilgan A. B. (1970) *Contemporary Scientific Psychology*, Academic Press.
Gross (ed.) (1971) *Handbook of Physiology*.
Grossman, S. P. (1966) 'The VMH: A Centre for Affective Reactions, etc.', *Physiology and Behaviour* **1**.
Hubel, D. H. and Wiesel, T. N. (1962) 'Receptive Fields . . . in the Cat's Visual Cortex', *Journal of Physiology*, **160**.
*Kane, A. (1971) Learning and Remembering', *Times Educational Supplement*, 17 September.
Lettvin, J. Y. *et al.* (1959) 'What the Frog's Eye Tells the Frog's Brain', *Proc. of IRE*, November.
Libet, B. (1965) 'Brain Stimulation . . .' in *Brain and Conscious Experience*, Springer-Verlag.
Miller, N. E. (1965) 'Chemical Coding of Behaviour in the Brain', *Science*, **148**.
Moruzzi, G. (1965) 'The Functional Significance of Sleep . . .' in *Brain and Conscious Experience*, ed. J. Eccles, Springer-Verlag.
Penfield, W. (1968) 'Engrams in the Human Brain', *Proc. Royal Soc. Med.* **61**.
Pitts, N. F. 'The Biochemistry of Anxiety', *Scientific American*.
Polani, P. E. (1967) 'Chromosome Anomalies and the Brain', *Guy's Hospital Reports*, **116**.
—— (1969) 'Autosomal Imbalance and its Syndromes, excluding Downs', *British Medical Bulletin*, **25**.
—— (1969) 'Abnormal Sex Chromosomes and Mental Disorder', *Nature*, **224**.

Smythies, J. R. (1970) *Brain Mechanisms and Behaviour*, Academic Press.
*Sperry, R. W. (1964) 'The Great Cerebral Commissure', *Scientific American*, **210**.
*Taylor, G. R. (1971) 'A New View of the Brain', *Encounter*, February.
*The Mind (1967) *Sunday Times*, 19 March.
*Walker, K. (1962) *Diagnosis of Man*, Pelican.
Young, J. Z. (1971) *An Introduction to the Study of Man*, Oxford University Press.

PART 3 CHEMISTRY AND TEMPERAMENT

Choh, Hao Li (1963) 'The ACTH Molecule', *Scientific American*, July.
Cobb, G. I. (1927) *The Glands of Destiny*, London.
*Cooper, Wendy (1971) 'Gender is a mutable point', *Daily Telegraph Magazine*, 10 December.
Davidson, E. H. (1965) 'Hormones and Genes', *Scientific American*, July.
Dawson, K. (1960) 'Effect of Menstruation on Schoolgirls' Weekly Work', *British Medical Journal*, 30 January.
——(1960) 'Menstruation and Accidents', *British Medical Journal*, 12 November.
*Funkenstein, D. H. (1955) 'The Physiology of Fear and Anger', *Scientific American*, **192**.
*Greene, R. (1971) *Human Hormones*, World University Library.
*Jost, A. D. (1970) 'Development of Sexual Characteristics', *Science Journal*, **66**.
*Kemble, J. (1969) 'Napoleon's Health', *Observer*, January.
*Klopper, A. (1970) 'The Reproductive Hormones', *Science Journal*, **66**.
Levine, Seymour (1966) 'Sex Differences in the Brain', *Scientific American*, **214**.
Loraine, J. A. et al. (1970) 'Endocrine Function in Male and Female Homosexuals', *British Medical Journal*, November.
*Mittwoch, V. (1971) 'Sex, Growth and Chromosomes', *New Scientist*, July.
Montgomery Hyde, H. (1970) *The Other Love*, Heinemann.
Morton et al. (1953), 'A Clinical Study of Premenstrual Tension', *American Journal Obstetrics and Gynaecology*, **65**.
*Mottram, V. H. (1946) *The Physical Basis of Personality*, Pelican.
*Pickford, Mary (1969) *The Central Role of the Hormones*, Oliver and Boyd.
*Shrewsbury, J. F. D. (1964) *The Plague of the Philistines*, Gollancz.
Spickett, S. G. (editor) (1967) *Endocrine Genetics*, Cambridge University Press.
*Walker, K. (1962) *Diagnosis of Man*, Pelican.
Whalen, R. E. (editor) (1967) *Hormones and Behaviour*, Van Nostrand.
Williams, R. H. (editor) (1968) *Textbook of Endocrinology*, Saunders.
Zuckermann, S. (1957) 'Hormones', *Scientific American*, March.

Index

ACTH (adrenocorticotrophic hormone) 154, 156, 166, 181
Abilities, different in man and woman 4, 43–6
  different in races 56–7
  teaching to maximise 61–4, 149
Accident proneness 75
Acromegaly 154
Adaptation 53, 145
Adoption in environment/genetic studies 19–20, 30, 31–3
Adrenal glands and hormones 42, 155, 158–9, 166
  see also Adrenaline and Steroid hormones
Adrenaline, balance with nor-adrenaline 158–61
  effect on synapse 93, 148
Adrenocorticotrophic hormone, see ACTH
Aggression, in car driving 84–7
  control, by hormones 159
    by limbic system 143, 145, 147
  and criminality 65, 77
Alcohol, effects of 112, 115, 148

Alerting centres 101, 125
Altman 109
Amygdala 142, 143, 145
Androgens, and intelligence 42
  maleness/femaleness due to 178
  production 159
  sex drive in women due to 175
  see also Testosterone
Anterior pituitary gland 154, 166
Anxiety, in animals 90
  in car driving 84–7, 89
  environment/genetic studies of 90–1
  limbic system control 146, 147
  neurosis 92–3
  survival value of 88–9, 93
Appearance and personality 22
Australian aborigine 56
Autonomic nervous system, branches of 159–61
  control by 'old brain' 100, 147, 160–161
  emotions controlled by 80–1, 83
  interacting with endocrine system 153
  overactive in anxiety neurosis 93

Average in nature 29

BABY LEARNING 87, 123–4
Balance, of behaviour, see Limbic system
 of hormones 161–3
 of internal environment 164–7
 see also Feed-back
Behaviour, aggressive, see Aggression
 control, external 5–6
  by drugs 147–8
  by electricity 142
  job satisfaction 2, 149
 control, internal, by limbic system 143–7
 criminal, see Criminal behaviour
 malnutrition affecting 54
 sexual 169–70, 175–7
 see also Personality
Brain, chromosome imbalance affecting, 102–7
 damage affecting 55, 57
 form and function 5, 80, 87, 98–101
 see also Cortex, Hypothalamus, Nerves etc.
 differentiating sex development 169, 178
 individuality of 43, 87
 and machine compared 113–15, 143, 150
 male and female compared 43
 memory and learning 87, 131–40
 protein requirements 54
 thyroid hormone affecting 157
 weight 109–10
Broadbent, D. E. 128–30
Bull, behaviour controlled electrically 142
Burt, Sir C. 15, 18, 29–30, 43, 47–9, 62, 63

CAFFEINE 148
Car driving, aggression and anxiety in 84–7
Cat, behaviour controlled electrically 142, 145
 vision 119, 121–2
 'wiring' pattern 124–5
Central nervous system 80
 see also Brain
Cerebellum, mongol 103
 normal 101, 110
Character, see Behaviour and Personality

Chemical, coding in memory 135
 energy required for brain function 112, 115
 experiments on behaviour 144, 146
 transmitter substances, in synapse 79, 111, 148
 in vision 117
 see also Drugs and Hormones
Childhood, prolonged in man 5
Children, IQ and education 61–4
 learning 81–2, 122–3
 see also Baby
 racial abilities 56–7
Chromosomes, effects of abnormalities 1–2, 20–2, 71–4, 102–4, 104–7
 nature and behaviour 6–13
 producing individuality 4–6
 and sex 38–42, 104–7
 see also Gene and Heredity
Coates, S. 21
Company, as environment 68–9, 84–7
Computer compared with brain 113–115, 130
Conditioned reflex 77–9
Conditioning, learning by 79–83, 87
Conscience 77, 81–3
Consciousness, state of 112, 131–2
 time lag before 118–19
Control of behaviour, external 5–6
  by drugs 147–8
  electrical 142
  job satisfaction 2, 149
 internal
  by limbic system 143–7
Corpus luteum 176, 179–81
Cortex, form and function of committed areas 98–101, 109–11, 133
 motor skill 132–3
 vision 116–22, 133
 uncommitted (interpretive) areas 100, 132–3
 memory and learning in 131, 135, 139–40
Cretin 170
Criminal behaviour, conditioning learning difficulty 79
 conscience lacking in 77
 environment/genetic influence on 65–70
 extra Y chromosome influencing 72–4
 and menstrual cycle 182–4
 produced in puppies 82–3

Cybernetic mechanism 115, 130

DNA (deoxyribonucleic acid) 3, 54, 163
Dalton, K. 182-3
Darlington, C. D. 43
Defects, children's recognized in teaching 62-4
  genetic, see Chromosome abnormalities and Gene defects
Delgado 142
Delinquency 1, 67
Dendrites 78, 110
  forming memory pictures 122-4
Deoxyribonucleic acid, see DNA
Depression 148
Deutsch, J. A. 138
Diabetes 159, 162-3, 165
Disease influencing behaviour 69, 172-175
Dogs, anxiety in breeds 90
  in conditioned reflex studies 77-9
  in criminal studies 82-3
  in genetic/personality studies 23
Drink control 144, 146
Drugs 22, 93, 147-8
Dwarfs 154, 174

ECCLES, J. 111, 119, 127
Education 59-64, 149
Electrical, control of behaviour 142-6
  impulses in nerve response 79, 111, 112, 117, 135
Elliott, H. C. 88, 147
Embryo, sex development in 167-70, 176, 178
Emotion, control by autonomic nervous system 80, 81, 83, 100, 118, 146-8
  and endocrine system 160-1
  in criminals 76, 84
  influencing intelligence 25
Endocrine system 151-71
Engrams 87-8, 128, 135
Environment, interacting with heredity, in anxiety neurosis 91
  in criminal behaviour 66-70, 73, 74
  in education 60-4
  in intelligence 29-33, 47-50
  to produce individuality 5-6, 14-15
  in racial IQ 51-8
  study methods 16-23
  womb as 13, 14, 55, 141, 177, 178
  'wiring' patterns formed by 125
Epilepsy 142
Extrovert personality 76, 77, 83
  and emotion combined in criminals 84
Eye 116-18
Eysenck, H. J. 76-7, 81

FSH (follicle stimulating hormone) 154, 156, 179-81
Family tree studies 17-18, 91
Fantz, C. D. 124
Fatigue 148
Feed-back, hormone 154, 156, 160, 166-7
  intelligence 54
  limbic system 145-6
Feeding behaviour 143-6
Female hormones 144, 154-5; 159, 175
  see also Oestrogens
Femaleness 38-9, 41, 106, 175-6, 178
  see also Woman
Finger-print, effect of extra Y chromosome on 105
Fisher, A. 144
Flashback in memory 134
Follicle (ovarian) 154, 179-81
  stimulating hormone, see FSH
Fraser Roberts, J. A. 20
Free-will 6, 65, 67, 150
Frog vision 119-20
Fröhlich's syndrome 146, 172

GH (growth hormone) 154, 156, 157, 158
Galton, F. 14, 17-19, 25, 36, 65, 91
Gene, in anxiety 90, 93
  basis of individuality 4-5, 11-15
  defective influencing, hormones 157-159
  IQ 20-1
  sex 170
  determining sex 8
  molecular effect of hormone on 163-164
  mutation 14, 53
  nature and behaviour of 8-13, 71
  see also Heredity
Genius 15, 17-18, 34-7, 38, 52
Giant 154

Glands, endocrine, link with nerves 160-1
  see Adrenal, Anterior pituitary, Follicle, Ovary, Pancreas, Parathyroid, Pituitary, Testes and Thyroid glands
Goitre 69, 157-8
Gonad 167-9
  see also Ovary and Testes
Gout 174
Gray, J. 42
Greene, R. 160-1, 162, 176
Gregory, R. 116
Grey matter, mongol 103
  normal 98, 110
  see also Neurone
Gross 121
Growth hormone, see GH

HALDANE, J. B. S. 64, 108
Health and personality 22
Hearing 101, 141
Heim, A. 44
Henry VIII 173-4
Heredity, interacting with environment,
  in anxiety neurosis 91
  in criminal behaviour 65-70, 73-76
  in education 60-4
  in intelligence 27, 29-33, 47-50
  producing individuality 4, 13-18
  in racial IQ 51-8
  study methods 16-23
  mechanism of 6-13
Hippocampal system 131, 134, 135, 142-3
Homosexuality 177
Hormones
  activity mechanism 163-4
  interacting with autonomic nervous system 159-63, 164-7, 178
  and learning 42, 170-1
  and menstrual cycle 179-84
  and personality 5, 172-5
  production 154-9
  and sex 42, 167-70, 175-9
Hubel, D. H. and Wiesel, T. N. 119
Hypothalamus, controlling autonomic nervous system 100, 142-8, 159-161
  controlling hormones 151-7, 162-7, 180

ICSH (interstitial cell stimulating hormone) 154
IQ, chromosome errors affecting 2, 20-1, 103-6
  and education 61
  and genius 35
  and malnutrition 54-5
  of mongols 103
  studies, in adopted children 19-20, 31-3
  in twins 30-1, 33
  tests of limited value 18-19, 25, 26, 35, 55-6, 60-1
  womb environment influencing 14
Idiocy 20
Individuality, of brain 43, 106-7
  'wiring' pattern 110, 124-6, 141, 151
  considered in education 61-4
  production of 4-6, 13-18
  of sexual behaviour 178-9
Insulin 159, 162-3
Intelligence, factors affecting 54-9
  see also Environment and Heredity
  feed-back 54
  male and female 38, 41-5
  of primitive man 53-4
  quotient, see IQ
  racial 26, 51-3
  sex chromosomes affecting 41
  tests for 18-19, 24-6
  see also IQ
Interstitial cell stimulus hormone, see ICSH
Introvert personality 76, 83

JACOBS, P. 72
Jensen, A. R. 51, 52, 61
Jervis, G. A. 20
Jewish ability 56
Jobs 2, 149
Jost, A. D. 168

KELLOG, R. 122
Klinefelter syndrome 105, 175
Knoll, M. and Kugler, J. 122

LH (luteinizing hormone) 154-6, 176-181
Lactic acid and anxiety 92-3
Lange, J. 66-70
Lead poisoning 55
Learning, baby 87, 123-4

involving central nervous system 80
  cortex 131-5, 138-40
  see also 'Wiring' patterns
  by conditioning 81-3
  and memory 128-31
  molecular change in 136
  and sleep 136
  thyroid hormone influencing 170
Lederberg, J. 55, 57
Lesbianism 177
Lettvin J. Y. 119
Libet, B. 118
Limbic system 100-1, 131, 134, 142-8
Loraine, J. A. 177
Luteinizing hormone, see LH

MacKay, D. M. 115
McWhirter, K. 73, 74
Male, hormone 144, 155-7, 159
  see Androgens
  sex chromosome, see Y chromosome
Maleness 41, 175-6, 178
  see also Man compared with woman
Man, chromosomes in 38-9
  compared with woman 4, 42-6, 84, 175-7
  intelligence developed in 53-4
  more than machine 6
Margolese 177
Marriage 18, 28, 69, 179
Meiosis 10-11
Memory, alphabet and pictures 122-124
  and learning 128-31
  molecular 135-6
  selective 129, 134, 135, 139
  and sleep 136
  storage 134, 135, 139, 150
  see also 'Wiring' patterns
  theories of 127-8
  trace 98, 100, 131, 139
Menstrual cycle 152, 175, 176, 179-81
  side effects 182-4
Mental, activity, no specific location in brain 101
  deficiency 20-1, 170-1
  brain in 43
  see also Cretin, Idiocy and Mongolism
  illness 21-2
    brain energy required 112
    see also Depression, Neurosis
Miller, N. E. 143

Mitosis 9, 10, 169
Mittwoch, V. 169
Molecular theory, of hormone activity 163-4
  of memory 135-6
Money, J. 42
Mongolism 22, 103
  anti-mongolism 104
Monkey, aggression controlled electrically 142, 145
  vision 121
Morruzzi, G. 136
Mosaic, chromosome effect 103, 104-5, 106
Motor skills, brain involved in 132, 133, 141
Mutation 14, 53
Myxoedema 174

Napoleon 3, 172-3
Negroes 26, 50-1, 54-8
Nerve cells 78, 80
  conditioned reflex involving 78-9
  in mentally defective brain 43
  web, see Engram
  see also Neurone
Neurone, abnormality in cretin 170
  associations, see Engrams and 'Wiring' patterns
  eye 116-18
  and hormone interaction 160-1, 170
  memory and learning involving 128, 130-1, 136, 138-40
  numbers 2, 110
    compared with computer tranistors 114
  structure and synaptic function 2, 80, 110-13
Neurosis, anxiety 84, 89-93
Newsom report 2, 43, 63
Nutrition, effects of poor 54-5, 141, 171

Obesity 146, 172, 173
Oestrogens 159, 177
  see also Oestradiol and Progesterone
Oestradiol 144, 159, 166, 180-1, 183
'Old' brain 100-1
  see also Limbic system
Orphans 19-20, 30, 31-3
Ovary, development 167-9
  in menstrual cycle 179-81
  not causing sex drive 175-6

PANCREAS 155, 159
Parasympathetic nervous system 80, 160, 178
Parathyroid glands 155, 158
Parry, M. 84
Pavlov, I. 77
Penfield, W. 133, 134, 135
Perception and brain 132, 142
Personality
  and appearance 22
  disease affecting 172–5
  and genius 35, 36
  environment/heredity effects, *see* Individuality and Behaviour
  hormones affecting 151, 158, 172, 174
  of mongols 103
  types 75–7
Phenylketonuria 20–1
Phosphenes 123
Pitts, N. F. 93
Pituitary, hormone control by 151, 154–7, 165–7
  malfunction 172
  menstrual cycle control 175, 180–181
  sexual behaviour control 169–70
Pleasure centres 143, 148
Pregnancy 54, 141, 181
Premenstrual symptoms 182–4
Progesterone 154, 159, 166, 180–1
Puberty 159, 177, 180

REM sleep 138
RNA, increased by hormone 163
  and memory 135–6
Rabbit studies 167–70
Race, differences in 4, 56, 57
  and IQ tests 26, 55–6
  intelligence, factors affecting 51, 55, 57
Rapid eye movement sleep, *see* REM sleep
Rat studies, aggression 142
  anxiety 90
  feeding 143–4, 145, 146
  growth 156
  learning 138–9
  protein deficiency 54
  sex 144–5, 167
Reflex, conditioned 77–9
Ribonucleic acid, *see* RNA
Rose, S. 136

SCHOOL, education theory in 57–64, 149
  as environment 14–15
  performance and menstrual cycle 182–3
  preparing for job 149
  sex bias in 43–4
Scribble and memory alphabet 122–3
Selective, breeding 47
  function of eye 111
  memory 129, 134, 135, 139
Sex, behaviour 144–5
  chromosomes 38–41, 167, 169–70
  imbalance effect 104–7, 170, 175
  development (embryonic) 41, 167–170
  drive 175–7
  hormones 152, 154–5, 159
  feed-back control 166, 167
  interacting with nervous system 175–7
  see also Hypothalamus
Sickle-cell anaemia 55
Sight, *see* Vision
Slater, Eliot and Shields, J. 21, 91
Sleep 112, 136–8
Smell 100–1, 133, 134
Smoking 148
Smythies, J. R. 134
Social, class and intelligence, of adopted children 20, 31–3, 47–50
  of races 26, 52–5
  implication of extra Y chromosome 73–4
Solomon 82–3
Spacial perception 41–2, 56, 106
Speech 98, 132, 142
Steroid hormones, feed-back control 166
  in menstrual cycle 152, 179–84
  see also Androgens
Stress 39, 89–90, 158
Sugar balance 159, 162–3
Suicide 184
Sympathetic nervous system 80, 159–160
Synapse, drug effect on 148
  mechanism 78, 79, 83, 111, 114, 124
  in memory and learning 135–9
  in vision 118

TSH (thyroid stimulating hormone) 155, 166
Talent 34–7

Tallness in men 22, 72, 154
Temperament, *see* Behaviour and Personality
Temporal lobe 99, 142–3
Terman, L. M. 36–7
Testes, development 167–9, 175
 as endocrine gland 155, 159
Testosterone, production 155, 159
 in rat sex behaviour 144
 in sex determination 169, 177–9
Thalamus 100, 118, 131
Thoday, J. M. 13, 60–1, 64
Thresholds in synaptic response 79, 83, 111, 114
 drug effect on 148
Thyroid, gland 155, 157–8
 malfunction 69, 157–8, 173–4
 hormone, feed-back control 165–6
 in learning 170–1
 stimulating hormone, *see* TSH
Trisomy 103–4
Turner's syndrome 106, 175
Twins in environment/genetic studies, of anxiety 90
 of criminal behaviour 66–70
 of emotionality 84
 of intelligence 29–31
 of personality 16–17, 21–2, 83

VERBAL REASONING 44, 56
Vision 101, 116–20, 133, 141

WOMAN, chromosomes in 38–9, 106
 compared with man, 4, 43–6, 84, 175–7
Womb as environment 13, 31, 55, 141, 171, 177–8
'Wiring' pattern in brain, defective in cretin 170
 individuality of 110, 125–6, 141, 151
 plasticity of 124
 and sleep 136, 138

Y chromosome, determining sex 2, 8, 12–13, 38–40, 167–9, 175
 effect of imbalance 22, 72–4, 105
Young, J. Z. 122